George Edwin Waring

Modern Methods of Sewage Disposal

For Towns, Public Institutions and Isolated Houses. Second Edition

George Edwin Waring

Modern Methods of Sewage Disposal
For Towns, Public Institutions and Isolated Houses. Second Edition

ISBN/EAN: 9783337804305

Printed in Europe, USA, Canada, Australia, Japan

Cover: Foto ©berggeist007 / pixelio.de

More available books at **www.hansebooks.com**

MODERN METHODS

OF

SEWAGE DISPOSAL

FOR TOWNS, PUBLIC INSTITUTIONS
AND ISOLATED HOUSES

BY

GEO. E. WARING, JR., M. INST. C. E.

SECOND EDITION, REVISED

NEW YORK
D. VAN NOSTRAND COMPANY

LONDON
SAMPSON LOW, MARSTON & COMPANY, LIMITED

1896

TYPOGRAPHY BY J. S. CUSHING & CO., BOSTON.

In the preparation of this book, it has been attempted to bring into convenient form, and within moderate limits, the more important results of the study and experience of many engineers, chemists, and biologists who have within the past quarter of a century exploited the various methods of sewage disposal. These results are set forth in a voluminous literature, and are clouded with much speculation and conflicting testimony.

It would not have been possible, even five years ago, to state with certainty some controlling principles which are now firmly established, and which are supported by the most successful practice of the past. That practice was based on empirical knowledge, and was the outgrowth of tentative effort. Little by little, it has been justified, elucidated, and corrected by scientific investigation, — notably at the hands of Schloessing, Frankland, Warington, and the corps of able men who have conducted the experiments of the Massachusetts Board of Health at Lawrence. These experiments were more complete than any that preceded them, and their outcome has been more practically conclusive: they must be of world-wide good effect.

In one sense, the art of sewage disposal is now only at the threshold of its real success, for only now can we work in the full light of day, divesting old processes of their defects, and devising new processes in accord with established knowledge.

While it is hoped that these pages will have some value and will offer some suggestion for engineers, they have been written with a view, as well, to the information of sewerage-committee men and others who may have occasion to look at the subject from the layman's point of view.

Especial attention is given, in the last two chapters, to the needs of isolated buildings, now so universally dependent on the noisome and death-dealing cesspool, which is, *facile princeps*, the great sanitary curse of the country.

The chapter on Chemical Treatment (XIV.) owes its arrangement and its completeness to my secretary and friend, Mr. G. Everett Hill, whose untiring research and discreet selection I desire most cordially to acknowledge.

G. E. W., Jr.

NEWPORT, R.I.,
April, 1894.

iv

CONTENTS.

———•◦•———

v

CHAPTER IX.

CHAPTER X.

CHAPTER XI.

CHAPTER XII.

CHAPTER XIII.

CHAPTER XIV.

CHAPTER XV.

CHAPTER XVI.

CHAPTER XVII.

CHAPTER XVIII.

MODERN METHODS OF SEWAGE DISPOSAL.

———o₀⦂o⦂₀o———

CHAPTER I.

GENERAL CONSIDERATION.

THE life of man involves both the production of food, directly or indirectly by the growth of plants, and the consumption and destruction of the organized products of such growth. The production and the destruction are constant. Between consumption and renewed growth there intervenes a process which prepares what we reject for the use of plants.

It is this intervening process that we have to consider in applying the comparatively new art of sewage disposal. The process itself has gone on from the beginning of the world, but it has been left to unguided natural action, which takes no account of the needs and conditions of modern communities.

In the primitive life of sparse populations, it was comparatively safe to disregard it; but, as population became more dense, and especially as men gathered into communities, it became increasingly important to bring it under control, for it then involved a serious

1   B

menace to the safety of the people. So long as our
offscourings could be scattered broadcast over the
ground, their destruction was attended with little danger; but when it became necessary to concentrate them
in underground receptacles, a capacity for real mischief
was developed. As these receptacles increased, with
the growth of communities, the menace increased, until,
in the light of modern knowledge as to the conditions
of healthful living, the need for radical measures of
relief became obvious. It is the application of these
measures that we are now to consider.

The sewerage of towns, and the drainage of important
buildings, are now controlled by expert engineers, and
they rarely fail to be reasonably well done. The economy of good plans is understood, and especially the
vital necessity for good construction. In fact, it may
be said that the adoption of excellent methods and
appliances for removing liquid wastes from houses and
towns is becoming general. It will in time become
universal.

This, however, is only the first step in sanitary
improvement. It is only the step of removal. It gets
our wastes out of our immediate neighborhood; it does
not destroy them. It is now recognized that quick and
complete removal is only the beginning of the necessary service, and that proper ultimate disposal is no
less important to health, to decency, and to public
comfort. The organic wastes of human life must be
finally and completely consumed. It is not enough to
get them out of the house and out of the town; until

they are resolved into their elements, their capacity for harm and for offence is not ended. It does not suffice to discharge them into a cesspool, nor does it always suffice to discharge them into a harbor, or into a water-course, leaving them there to the slow processes of putrefaction.

The need for improving the conditions of sewage disposal has long been recognized, and, especially in connection with large foreign towns, efforts of the most costly character have been made to obviate accumulations due to the discharge of sewers. The floods made foul with the wastes of the huge population of London have been poured into the Thames, until, in spite of years of effort to relieve that river, its condition has become, in the language of Lord Bramwell, "a disgrace to the Metropolis and to civilization." The millions expended since 1850 on the still unsolved problem have not thus far effected more than a mitigation of the evil. London is to-day, apparently, as far as ever from its ultimate solution, though of course the former direct discharge of sewage all along the river front, and the resulting local stench, have been suppressed. The case grows in gravity with the growth of the population, and measures which promise success when adopted are not able to cope with the greater volumes produced later. While substantial relief has been secured in the case of other towns in England and on the continent of Europe, there is not generally as yet such an early and complete reduction of organic wastes, without offensive putre-faction, as the best sanitary condition demands.

In our own country, New York City and the towns on the Mississippi, and on other very large rivers, have such tidal and flood conditions as to secure satisfactory disposal by dilution and removal. At Boston, Philadelphia, and Chicago, the needed relief can, under the methods adopted, be secured only by works of the greatest magnitude and cost, while the smaller towns have, as a rule, yet to devise methods by which, unless they are exceptionally well placed, they can destroy their wastes at a practicable cost. The importance of relief is being more and more realized, but the means of relief are little understood by the people. A wider appreciation of the efficiency of these means is a necessary condition precedent to general improvement.

Systematic works, chiefly by removal through intercepting sewers, have, until recently, been confined to cities. Smaller towns are now perfecting their methods of removal, and there is a growing desire to find means for purifying the outflow which will not cost more than can be afforded. Interest is also growing among householders, who are becoming convinced of the dangers of cesspools, with their retention of putrefying wastes within contaminating reach of houses and of their sources of water-supply.

In its progress thus far, the art of disposal has worked itself out mainly by progressive practice. It began in the instinctive desire to get offensive matters out of sight. As new difficulties presented themselves, and as the requirements of a better civilization arose,

new methods were devised for better concealment in the
ground, or better removal by sewers. In fact, such
concealment by the use of vaults and cesspools, and
removal by large sewers, still remains, with the majority
of the people, the accepted method of dealing with the
more obvious difficulty of the situation. The difficulties
which are not so obvious, the serious ultimate difficulties, these methods fail to relieve. It is hardly half a
century since the dangers of incomplete sewage removal
were appreciated and radical measures of relief were
attempted. In London, large brick sewers, not only in
the streets, but under and about houses, which had long
existed as the seats of foul deposits, now had their condition pointed out, and a "Blue Book" of the British
Parliament, published in 1852, set it forth in a manner
to secure effective attention. It was shown that these
sewers and drains were so large that they could not be
kept clean by their natural flow. It was then that the
movement for the use of pipes for sewerage and house
drainage received its first great impetus.

In 1857 there was presented to Parliament a report
by Henry Austin, C.E., on the means of deodorizing
and utilizing the sewage of towns. There followed,
not only an improvement of much of the local drainage
of London, but the carrying out of a plan for keeping
foul sewage out of the Thames, by collecting it in
reservoirs some miles down the river, to be discharged
at the beginning of the outgoing tide. As has already
been intimated, this work was ineffective, so far as the
main purpose of purifying the Thames was concerned,

and the problem seems still to be overtaxing the capacity of English engineers.

There was, at that time, little knowledge of the proper means of relief in such cases, and the enormous sums spent in works for the discharge of the sewage on the outgoing tide soon proved to have been a misdirected expenditure. The art of sewerage had, for many years, confined itself to an improvement of the means for distant removal, and the world accepted and still accepts, as a part of the policy of its great cities, the inevitable construction of majestic and costly engineering works for this service, carrying not only foul sewage, but floods of storm water as well. It is now demonstrated that, even at London, and in all but a few exceptional conditions, like those of New York, where the whole harbor is flushed twice a day by the great tides coming in through Long Island Sound and passing out at Sandy Hook, and of towns on great rivers, the effect of such works is, largely, to remove the point of deposit, not to prevent deposit, and that the great volume of their discharge has often added to the difficulty of final disposal.

The lower Thames has reverted to a condition which is said to be hardly better than that of forty years ago. As a general rule, wherever a copious discharge of unpurified sewage is made into a river or harbor having insufficient volume of flow, the condition grows worse as the population increases, and as a heavier duty is imposed on a limited capacity for dilution. In many cases the difficulty has become so serious that those who are charged with its consideration see that it presents one of the most troublesome problems with which municipal authorities are confronted. Sooner or later, the provision of some means of purification, or, at least, of the removal of the grosser impurities of the sewage, becomes imperative, and the question of sewage disposal is assuming greater importance year by year.

The tendency of legislation, here as well as abroad, is toward the prohibition of the fouling of rivers, thus far mainly for the protection of sources of water-supply. This is doing much, and promises to do more, in the way of restricting the free discharge of sewage into streams. There is also a growing sentiment in favor of cleanliness, and causes of offence which have hitherto been disregarded are now attracting attention. Those who occupy lands past which streams flow are beginning to assert and to enforce their undoubted right to have them flow in their natural unfouled condition.

The adoption of measures for the purification of sewage by some English towns has been compelled by the firm and lawful demand of individual land-owners below them; and here, as well as there, the necessity for purification assumes increasing importance. This is especially true in the case of towns situated on the minor streams. While these are natural and sufficient drainage outlets, so far as storm water is concerned, they are often insufficient properly to dilute the sewage sent into them. These small towns, many of them having mere brooks for outlets, are growing rapidly, and modern methods of drainage

are fast leading to the more and more complete discharge of water-borne filth by sewers.

So, too, on the larger streams, villages are growing
to towns, towns are growing to important cities, and
conditions which were formerly tolerable are now becoming intolerable. The Schuylkill River, for example,
which is the most important source of water-supply for
Philadelphia, is lined with populous and growing manufacturing towns, which have only this river for an
outlet, and which also take their water-supply from it.
The same conditions exist along many of the rivers of
New England, and throughout the older parts of the
country generally, and they are extending westward.
It is therefore clear that, in the case of towns not lying
on the larger rivers, public sentiment and the rights of
riparian owners will demand the increasing adoption of
means for withholding crude sewage from them.

The following was written in 1888: "It is not likely
that towns situated on great rivers or on the seacoast
will, for a long time to come, give thought to any other
disposal of their sewage than its discharge directly into
the river or into the sea.   As the country fills up, and
as towns situated on small streams, or on no stream,
increase in size and in wisdom, they must, perforce,
seek for some means to get rid of the copious flow of
water, made foul by its passage through the houses and
shops of the people. The indications are clear that
legislative control of this matter cannot long be delayed,
and there is no more intricate or more interesting problem now presented to the sanitarian than the correct

solution of this great question of the future. Its final
solution implies a better acquaintance with the ultimate methods of organic decomposition and filtration
than any one now possesses. It seems, however, as
though the scientific world had at last reached the
threshold of real knowledge concerning the processes
by which organic matter is converted into those mineral
compounds which, inoffensive and innoxious in themselves, become, in the economy of life, the direct food
of growing plants. It is these processes that we must
employ in the successful destruction of all organic
waste other than such as is consumed by fire. They
go on in spite of us; we may delay them, or conceal
them, or change the seat of their activity; we may
hasten them, or modify them, but we cannot prevent
them. Sooner or later, by combustion, by direct putrefaction, or by indirect fermentation, they will work
their destructive end, bringing all matter that has once
lived again back to the domain of life. The cycle is
unceasing, and according to our action concerning it,
or according to our neglect, will its influence be good
or bad. Thus far, we are not quite sure how our action
should be guided." [1]

During the years that have since elapsed, the most
important investigations of the Massachusetts State
Board of Health, carried on at the experiment station
at Lawrence, under the direction of Mr. Hiram F. Mills,
C.E., have confirmed the theories then held, and have
thrown much light on the methods by which they may

[1] Sewerage and Land Drainage, Waring, p. 233.

best be reduced to practice. It was only after this clear definition and demonstration of the processes involved, and of the methods of their application, that we were in a position to work with real knowledge. Then only could empiricism be made to give place to well-established theory.

Could we now set aside the influence of long years of practical work, the atmosphere would be greatly cleared; but practical work has a very persistent influence, and the art of purifying sewage will long feel the effect of experience with methods which would not have been devised in the light of what is now known. When the first attempts were made to get rid of the impurities of sewage by artificial means, great importance was universally attached to their manurial value, and promise was held forth of great profit to result from their development in a useful form. The obstacle of extreme dilution was not appreciated, and it was long before the discovery was made that, as with the gold said to exist in sea-water, the attempt to separate these matters by artificial methods would cost more than they were worth.

The belief also prevailed that the chief source of offensiveness of sewage lay in the solid fæcal matter that it contained, and this belief still finds much popular acceptance. One of the most prominent sanitary exhibits at the World's Fair in Chicago, a Russian invention, has the separation of this matter for its chief end, and the description accompanying it urges such separation as the *sine qua non* and the chief need of hygienic improvement. Even in Paris, where the purification of sewage is being carried out on a very large scale, and where its requirements are well understood, the use of the *tinette filtre*, which holds back the solid parts of house drainage until they putrefy, and then allows them to flow to the sewer, in this worst possible form, has within recent years been allowed to come into extensive use. At Newport the old rule still prevails largely, that house drainage shall be retained in cesspools until it can, after decomposition, overflow as a foul liquid into the public sewers. The fact is, that fæcal matter is of far less consequence than urine and the waste of the kitchen sink.

Then, too, it was long thought that if sewage could be purged of its suspended matter,—of that which clouds it and colors it,— purification would be effected. An imperfect clarification by mechanical or chemical processes is still applied in some cases where a high degree of purification is really needed, although it is now well known that such clarification does not and cannot remove from sewage its most putrescible matters, nor its minute living organisms. Imperfect results, which have satisfied legal requirements in Europe, are in such cases accepted as sufficient, in spite of a recognition of their incompleteness.

The purification of sewage is surely on the eve of great extension in this country, and it is necessary to its success that the importance of making it as thorough as possible be made known, as well as its conditions and requirements. If the work is to be done at all, it

is surely worth while to do it well. Half-way meas
ures, like chemical precipitation, may satisfy present
legal demands, and they may, in exceptional cases, be
advisable, but they will not meet the requirements of
the better-informed public opinion that is now growing up. The means for entire purification are within
reach, and imperfect results will not long be accepted
as sufficient.

In practical work, two cardinal principles should be
kept in view, and should control our action : —

(*a*) *Organic wastes must be discharged at the sewer outlet in their fresh condition,— before putrefaction has set
in ;* and

(*b*) *They must be reduced to a state of complete oxidation without the intervention of dangerous or offensive
decomposition.*

In considering the requirements of disposal, we
should not be too much influenced by the experience
and practice of England, where a natural tendency to
conservatism has led to the continued use of methods
which would not have been adopted had engineers
known twenty years ago what is well known now. It
is not unusual to see it stated in discussions of this
subject, that English experience points to the provision
of one acre of sewage irrigation land for each hundred
of the population, and projects for irrigation are sometimes rejected because of the great cost that would thus
be entailed. The fact is, that no such great amount of
land is required for the proper treatment of the foul
elements of the sewage. If needed at all, it is needed

only to meet the demand of great floods of storm water, sent to the fields at a time when they are already saturated with rain.

This difficulty is sometimes insuperable, in connection with the use of sewers to which storm water from the whole town has access, and such cases must be considered by themselves. Where all storm water is excluded from the sewers, this difficulty does not exist, and the irrigation area may be limited to an amount that will, when wet with rain, still admit the volume necessarily resulting from the copious use of water in our houses. Just what this limit is to be, cannot now be determined. It will vary according to the character of the soil. Clay and peat will absorb much less water than loam, sand, and gravel will, and ground underlaid with a porous subsoil, or thoroughly underdrained, will absorb more than ground underlaid by undrained, compact clay. At the same time, if only heavy land is available, this may be much improved by artificial drainage, and the freedom of its drainage will increase as time goes on, so that it will ultimately be able to cope with an increased flow.

In all forms of irrigation and filtration disposal, intermittent application is the key to success. Completeness of purification may be favored by long intervals, but capacity for purifying large volumes calls rather for intervals as short as will suffice to get rid of surface accumulations and to maintain a pure effluent.

The oxidizing organisms are short lived. When their food has been consumed, they disappear, and their

disappearance implies a reduction of oxidizing capacity.
This can be maintained at its maximum only by keeping up the full working force that the soil we use can
accommodate. The condition in this respect cannot,
of course, be regulated with anything like exactness,
but experience and increasing knowledge will enable us
so to adjust the supply to the demands of full activity
as to approach more and more nearly to a maximum
efficiency. The indications are, that the conditions
affecting absorption, aeration, and the maintenance of
a full supply of organisms will improve with use, and
especially with frequent use; so that the capacity of any
suitable soil to purify sewage may be increased, not
indefinitely, but to a point that has nowhere yet been
reached, at least in irrigation.

We are still far from knowing enough of the detailed
working of disposal processes, to determine what would
secure the best possible result in any given case, but we
are at least so sure of our ground that we can work in
the right direction; and we can now safely cut loose
from the restrictions imposed by early English practice,
where, in view of the occasional great increase of flow
over a rain-saturated soil, due to the admission of storm
water to sewers, and of the supposed need for maintaining a good agricultural condition, it was deemed necessary to provide very large areas of land, and where it
was thought that, even then, considerable periods of
rest were requisite.

Under conditions which are often available, we may
feel safe with an acre for each five hundred of the popu-

lation, and under conditions which are hardly exceptional, one acre per thousand of population may suffice. This was frequently reached in the Intermittent Downward Filtration Works of J. Bailey Denton, in England.

At Lawrence, volumes of sewage equal to over 100,-000 gallons per acre per day were continuously made purer than the best drinking-water supplies.[1] When, as is usual, it is only a question of making the effluent as pure as the water of ordinary streams, much larger volumes may be constantly taken care of.

The practical results of broad irrigation in works of long standing show that the process, when well carried on, is devoid of all offence, and may be made to yield agricultural returns, which will go far toward paying the cost of maintenance. At Gennevilliers, where irrigation and filtration are combined, and where the soil is gravelly, the sewage of Paris is made purer than the best drinking-water of that city. Prosperity has been brought into the district, which was originally a poor one. Land that was formerly of trifling value now sells for nearly $1000 per acre, and its rental value has quadrupled. The population has increased one-half, and general prosperity has taken the place of comparative penury. The health of the people is excellent, and even in 1882, when there was a cruel

[1] "Varying quantities of sewage were applied to this tank during 1890. At the end of the third year the tank was filtering about twice as much as at the end of the second year, with a rather purer effluent. The amount was about 102,000 gallons per acre per day. The better result is attributed to raking over the surface to a depth of one inch once a week." — *Mass. Report.*

epidemic of typhoid fever in Paris, there was none here. The general mortality of Gennevilliers in 1865 was 32 per thousand. In 1876 it was 25, and in 1881, only 22. Measures have been taken to extend the same system over other lands, sufficient for the purification of the entire outflow of the sewers of Paris, save during floods of the Seine, when there is no objection to its direct discharge therein.

Correspondingly good results have been reached at Berlin, Breslau, and elsewhere. In fact, the entire adequacy of disposal by application to agricultural lands has been fully and finally demonstrated.

Wherever suitable land can be had, this method of disposal meets with no obstacles which experience has not shown us how to surmount, and it encounters no prejudices which acquaintance with its details does not remove. Sewerage engineers are sometimes asked where it is best to "dump the sewage," and other expressions are used, suggestive of the disposal of putrid filth by night-soil carts. Sometimes the owners of adjoining lands object to the establishment of a sewage field, in the belief that it will become a nuisance.

The fact is, that the "dumped" matter is fresh and inoffensive, is mainly invisible, save as it clouds the flow, and is a thousand times diluted. If its treatment is properly regulated, it is withdrawn from the diluting water and completely destroyed, in a manner that is imperceptible to our senses, and, in the case of broad irrigation, with the effect of producing a luxuriant vegetation during the growing season. At Wayne,

Penn., a protesting neighbor, who had apprehended an insufferable nuisance from the disposal field, soon expressed a regret that his land was not so situated that the sewage could be made to flow over it.

If there is still room for doubt on any point, it is as to the character of the few bacteria which escape the action of the process employed and are found in the effluent. It is not known that disease germs ever exist among these, and it is altogether probable that they do not. So far as these organisms are understood, it is thought that they cannot withstand the destructive activity of the oxidizing and nitrifying organisms which are always present, and it is believed that only these hardier organisms exist in the effluent of land-purification works. Certain it is that no instance has been reported where contagion was carried by such effluents, and experience at Gennevilliers has shown that typhoid fever and cholera, when rife in Paris, were completely arrested at the irrigation fields. At Berlin, also, it is stated that no disease has been suspected of having been communicated by the sewage to any one of the large population of the irrigation farms.

In the Massachusetts table of comparison of the purified effluent of seven sewage filters and the waters of seven wells used for drinking by many persons, it is shown that there were three and a half times as many bacteria in the well waters as in the effluents. It is perfectly safe to say that, of all methods thus far devised for purifying sewage, its proper application to land has secured much the most complete destruction of bacteria, and that it has never given occasion to suspect the survival of disease germs.

While holding the opinion that there are compara-tively few cases in which some form of application to land is not better than any process of chemical precipi-tation, it is not pretended that the latter is never desir-able. At Worcester, Mass., for example, where the requirements of purification could have been satisfied more completely and at less cost,—however low the degree of clarification demanded in that case,—there was another and a controlling factor that indicated the use of the chemical process. The streams flowing through the city, and so fouled by its sewage as to demand treatment, were tributary to Blackstone River. While Worcester might use the water as it pleased, it was obliged to pass it on, practically undiminished in volume, to mill-owners, who used it at successive dams along its lower course to Narragansett Bay. To flow over land such an immense volume as these fouled tributaries delivered, or to pass it through purification channels, would have led to such an amount of evapora-tion, especially at the season when the river is naturally at its lowest stage, as might lay the city open to endless suits for damages. This is doubtless the consideration which led to the adoption of chemical treatment in this case.

Therefore, while chemical processes may properly be relegated to a very secondary position, as being too costly and too ineffective for general use, they are by no means to be considered as worthless.

In the following chapters, the attempt will be made to set forth, in a form available for practical use, the best methods of applying the principles on which sewage purification rests, having due regard to the teachings, but following with great caution some of the traditions, of the past experience of the world in efforts at sewage disposal.

CHAPTER II.

WHAT SEWAGE IS.

ROUGHLY speaking, all the water-borne wastes of a community go to make up its sewage. More accurately, the term is applied to what is removed by underground channels, or sewers, — as from a town to a river, or from a house or public institution to a cesspool, or other outlet. It is not usually applied to gutter flow, as such, though where there are storm-water sewers, this becomes, during rains, by far the most considerable, and in some regards the most troublesome, element of sewage.

The sewage of different towns is reasonably uniform in composition, save where the sewers receive the discharge of woollen mills, dyeing establishments, and other industrial works which produce a large effluent, often surcharged with chemicals and coloring-matters. This factory waste usually may be, and where practicable it always should be, cared for by those who produce it. Such care may more often than not be made to pay more than its own cost, in the value of the materials recovered. When this waste must, of necessity, enter the sewers, it may modify the treatment required in disposal.

20

Setting this element aside, sewage may be divided into two distinct classes: —

a. That delivered by the separate system of sewerage.

b. That delivered by the combined system.

a. The separate system receives only water that has been made foul in the houses and shops of the people, more or less diluted with ground water leaching into the sewers, or specially collected by underdrains and discharged with their flow.

Its volume is substantially that of the water-supply of the community, plus the ground water. The waste matter which this water receives in its use amounts, comparatively, to very little. Mr. Mills places it at two parts per thousand, one of these parts being mineral matter, which need not now be considered, and the other (about four tons per million gallons) being organic matter, to remove and destroy which is the purpose of works of "disposal."

The chief source of this matter is the food prepared for the use of the people, and either rejected in the course of preparation or actually consumed. To this are to be added the soiling of linen and the soap used in washing and bathing, and the washings of dishes and kitchen utensils. Coarse garbage and swill are not included. A certain amount of industrial waste is also to be considered, even where there are no factories; the washing of stores, market-stalls, etc., and the drainage of stables have a certain importance. These may be regarded as inevitable and constant elements of all town sewage.

Two of these elements are much more important than the others. They are the excretions of the people and the waste of kitchen and pantry sinks. These are the greatest in quantity and the most serious in quality. From the point of view of the sanitarian, they are about equally important. There are no statistics as to their relative volume, but it is generally assumed that the sink waste is more abundant than the excretions.

Both consist mainly of elements of the food-supply. In one case they have, and in the other they have not, been eaten and digested. The difference between them is a difference in condition, rather than in constituents. One is further advanced in decomposition than the other, but both are moving along the same line, and they will soon become indistinguishable. A cask of water-closet matter and a cask of kitchen-sink matter will, after a short time, be the same in appearance, in odor, and in constituents. They will be equally offensive and equally dangerous, save that the closet matter may have been originally infected with disease germs, —such as the typhoid bacillus,—and the other not infected. Even this distinction is not likely to continue in any practicable treatment of the two. They must therefore be regarded as demanding, in all respects, the same care and precautions.

It is a favorite idea of theoretical sanitarians, especially during the early stages of their professional development, that danger to health lies chiefly in that single element of our wastes which has a foul odor when first produced, and which is, at times, infected with the bacilli of cholera and other diseases of the digestive organs. Even so clever and practical a man as the late Fleeming Jenkin, F.R.S., Professor of Engineering in the University of Edinburgh, published, some fifteen years ago, a project for establishing two distinct systems of piping in every house, for the separation of fæcal matter from other wastes. Professor Jenkin's lectures were edited for publication here by the present writer, with no other qualifying remark as to this theory than that, "with the best American workmanship, the constant American practice of making the two systems in one is entirely safe, and its simplicity and economy are commended."

The Liernur system of sewerage is based on this idea, that danger lies solely in fæcal matter, and on the notion that this contains the chief manurial value of our excretæ. A very ingenious Russian invention, shown at the Columbian Exposition, had, as already stated, for its chief purpose the separation of the more insoluble portion of the fæces from its soluble portion, and from urine and other liquids.

All this is gross fallacy. While fæces are originally most offensive, and while they are, as has been said, subject to a certain class of infection before they leave the body, they are, as an element of sewage, relatively unimportant in amount, and their infection may be instantly communicated to other wastes with which they come in contact. Their odor also is ephemeral. It is due to an exhalation of gases, which is suppressed as soon as they are submerged.

Dr. C. Meymott Tidy presented to the Society of Arts (London) in 1886,[1] a paper on "The Treatment of Sewage," in which he went fully into the relative importance of different elements of sewage, beginning as follows: —

"*Liquid Excreta.* Every adult male person voids on an average 60 ounces (3 pints) of urine daily. The 60 ounces contains an average of 2.53 ounces of dry solid matter, consisting of —

Urea	512.4 grains.
Extractives (pigment, mucus, uric acid) . . .	169.5 "
Salts (chiefly chlorides of sodium and potassium),	425.0 "
	1106.9 "
	= 2.53 ounces."

The urine, therefore, of a population of 10,000 adults may be taken as 600,000 fluid ounces, or 3750 gallons per day.

"Urine rapidly decomposes, the urea becoming the volatile body, carbonate of ammonia, and the urine thereby losing a valuable manurial constituent. After a time, but at a later stage, certain foul-smelling gaseous products of decomposition are evolved. To collect and preserve urine, therefore, presents practical difficulties. The ammonia from stale urine was formerly distilled and converted into a sulphate, at Courbeville near Paris.

"*Solid Excreta.* Every adult male person voids about 1750 grains (or 4 ounces) of fæces daily, of which 7

[1] Journal of the Society of Arts, Oct. 8th, 1886, p. 1127.

per cent is moisture. The dry fæcal matter passed daily is therefore about 1 ounce per adult head of the population. Of this *dry* fæcal matter, about 88 per cent is organic matter (of which six parts are nitrogen) and 12 per cent inorganic, of which four parts are phosphoric acid. Of this dry fæcal matter, 11 per cent is soluble in water.

"Taking a population of 10,000 adults, it follows that the moist fæcal matter passed daily is equal to 2500 pounds (1 T. 2 cwt. 8 lbs.) or 1.116 ton, whilst the dry fæcal matter is equal to 625 pounds (5 cwt. 2 qrs. 9 lbs.).

"The facts, therefore, respecting the excreta of a population of 10,000 adults may be thus tabulated: —

TABLE I.

Fæcal Matter passed per 10,000 of Adult Population per diem.

Moist fæcal matter excreted . . . . 2500 lbs.
Dry fæcal matter excreted (calculating 75 % as moisture) . 625 "

$$\left.\begin{array}{l}\text{Soluble in water} \quad = \quad 68.55 \text{ lbs.} \\ \text{Insoluble in water} = 556.45 \text{ "}\end{array}\right\} = 625.$$

TABLE II.

Urine and Fæces passed per day by 10,000 Adults.

	Total Solids. Moist. Lbs.	Water. Gallons.	Solids. Dry. Lbs.	Solids. Soluble. Lbs.	Solids. Insoluble. Lbs.
Fæces	2500	187.5	625.0	68.55	556.45
Urine	. . .	3750.0	1581.21	1581.21	. . .
		3937.5	2206.21	1649.76	556.45

"The table on page 1128 [omitted here] has been adapted from Letheby. The quantities given are somewhat below the normal. The facts were collected from a number of sources, the ratio of children to adults being that adopted by Roderer and Eichhorn.

"My own experiments would lead me to give one pint as an average quantity of urine passed by children daily, up to the age of ten years, the quantity gradually increasing up to three pints in the adult. The solid constituents of the urine which, at the age of ten, are on an average 0.8 ounces daily, increase, according to my observation, up to 2.5 ounces in the adult. The quantity passed by girls and women is rather less than that passed by boys and men.

"The fæces passed by girls and women are considerably less than that passed by boys and men. The calculations in the table state the amount as less than one-third. My own observations, however, scarcely support these numbers. It would, I think, be more accurate to regard the fæcal matters passed by female children and adults as about one-half that passed by male children and adults."

The general showing of the omitted table is to the effect that, while there is two and a half times as much *solid matter* in the urine as in the fæces voided by an adult man, there is more than two and a quarter times as much in the voidings of persons of all ages and both sexes, when taken in the proportion in which they ordinarily exist in town populations.

In these wastes, as in all others, the pure water contained may be entirely disregarded. Attention is demanded only by the solids associated with it, and the solid parts of the fæces are not more than one-third of the total solids of the voidings. The solids contained in elements of sewage due to other sources — of an equally troublesome character, and ultimately of equal manurial value — still further reduce its proportion. We must, therefore, without in any wise belittling the importance of fæcal matter as an element to be dealt with, give even greater attention to other wastes, which have too often been considered as secondary.

There is, of course, in all domestic sewage, a large amount of woody fibre and other non-nitrogenous material, which by itself is neither offensive nor dangerous; but diffused through the mass, and made difficult because of it, there is a large amount of other material. It is to this that the work of purification is really directed.

We see, therefore, that the outflow of a system of separate sewers contains, borne in a vast volume of water, much comminuted rubbish of an innoxious character, and intimately associated with it, other matters containing nitrogen, which may be infected with minute organisms capable of conveying disease, and which may, in any case, give rise to a serious nuisance, contaminating air, water, and soil, and inimical to the sense of decency, to the comfort, and to the health of the people.

This dangerous element, with the less objectionable solids accompanying it, is to be separated from the water which serves as its vehicle and subjected to processes which will destroy it.

It is a great, perhaps the chief, merit of a properly constructed separate system of sewerage, that it delivers all of these wastes at the sewer outlet in a fresh condition, — in a condition in which they might flow through a perfectly smooth and well-washed street gutter without attracting attention by their odor.[1]

It is still another very great merit, especially from the point of view of sewage disposal, that the volume of outflow to be treated is substantially uniform, day by day, and week by week, so that it may be treated in a uniform manner by a uniform plant.

b. The combined system receives all of the above-described wastes, and, in addition, it takes in the intermittent and sometimes copious flow of storm water falling on the town.

This surface flow offers, from the point of view of the subsequent artificial treatment of the discharge, three serious difficulties : —

1. It interrupts the uniformity of the volume to be treated, sometimes increasing it by as much as fifty-fold, and this at a time when irrigation fields and filtration areas are encumbered with the water of heavy rains falling directly upon them. Even the effort to obviate this by the use of storm overflows only mitigates the condition, for such overflows must be placed so high in the sewer as not to discharge any part of the most

---

[1] It is assumed that such abominations as large " pot " traps, catch-basins, and grease-traps will be prohibited. These are all devices for retaining what should flow on uninterruptedly to the outlet of the sewer, and are seats of foul decomposition. They do no good that may not be accomplished otherwise, and they do much harm.

extreme dry-weather flow, provided for in view of the future growth of population. During even light rains, therefore, the sewer will carry continuously a volume larger than the extreme estimate for the moment of greatest use of a remote future.

For example, if we assume the case of a growing town of 5000 inhabitants, with a daily consumption of 60 gallons per head, or a total of 300,000 gallons, it would not be unreasonable to provide for a future population of 25,000, and, in view of possible future contingencies, to place the ultimate water consumption at 100 gallons per head, or a total of 2,500,000 gallons. On this basis, the overflow would have to be put high enough to allow an extreme discharge of about 3000 gallons per minute, with a good margin to spare. Therefore the constant flow, during any considerable rain, would be in the neighborhood of 5,000,000 gallons (allowing for margin) per day, against a dry-weather flow of 300,000 gallons now, and a possible 2,500,000 gallons ultimately. This would require an irrigation area, or a filtration or chemical plant, eight times as large as the present dry-weather requirement, and twice as large as the possible, but not probable, ultimate requirement.[1]

2. The strong flow reaching the sewers from street gutters is not mere diluting water. It comes laden with gravel, sand, silt, horse droppings, leaves, paper,

---

[1] Of course it would be possible to modify this condition by starting with a low overflow and raising it as population increases, but even then the increase volumed due to rainfall would be vastly beyond the volume of the dry-weather flow.

and various rubbish, which not only overcharge the screens at the disposal works, but add immensely to the sludge of chemical precipitation tanks, and greatly embarrass filtration and irrigation.

3. While the sewage delivered by the separate system under consideration is all in a fresh condition, that from the combined system is already more or less advanced in decomposition, due not only to the putridity of the retained contents of the inlet catch-basins, which is washed into the sewers by each rain, but also to the accumulation of street rubbish in the sewers themselves, which catches fæces and other organic wastes, and holds them until they putrify. The next storm may wash them all forward to the disposal works, but it will bring in a fresh supply of rubbish that will soon renew the bad condition. This last difficulty may be obviated by proper construction and proper flushing, but it is widely prevalent in sewers of this system.

The considerations above set forth have led to a general acceptance of the conclusion that, when sewage has to be pumped, and when it has to be subjected to artificial methods of cleansing, the separate system is to be recommended.

With this system the problem of disposal is reduced to its simplest form, and is susceptible of the easiest and best-regulated solution.

What we then have to deal with is fresh organic matter, of reasonably uniform composition, mainly reduced to a state of fine comminution, and suspended or dissolved in water of nearly equal daily volume.

CHAPTER III.

THE end sought in sewage purification is to separate the one part of organic matter from the thousand parts of water, in such a way as to leave the water pure and to render the organic matter harmless. If we do this, we satisfy the sanitary requirement.

The restoration to the domain of life of that which we have discarded is only an inevitable natural process, which we may be able to hasten, and which we may turn more or less to our advantage, according to our skill in developing its agricultural possibilities. The restoration will, in any case, go on; Nature, in her rough way, will take care of that. What we have to do is so to guide her processes as to prevent annoyance and harm to the people, and to derive what benefit we may by converting them into processes of fertilization.

If we so far secure the resolution of this matter into its elements as to make it available for plant food, the work of purification is accomplished. Whether or not we use it as plant food, is of no direct consequence from a sanitary point of view. This is a purely economical question, to be decided according to circumstances. The capacity of the irrigation field or of

31

the filter to remove the wastes from the flow will be
aided by vegetation,—both by increased evaporation,
and by the great aid to absorption due to the presence
in the soil of roots and of the channels left by dead
roots,—and vegetation may generally be made to yield
a useful return. Purification, however, will be complete without this, and the absorptive condition of the
land will be as much improved by the growth of weeds
which are to be wasted, as of crops which are to be used.

The complete purification of the water of the sewage,
—its conversion to the condition of the best drinking-
water,—and the reduction of its impurities to their
elemental forms, are entirely within our reach, but it is
not, in all cases, worth while to carry our work so far.
It may be necessary that it be continued only to a given
point of partial purification, and where this will suffice,
the cost of further treatment may be avoided.

It may be a question only of withholding such obvious impurities as cause decided discoloration, such
of the coarser parts of the sewage as would become
stranded on a shore, and such as would form deposits in
a stream or harbor. Or it may be required that the
sewage be brought only to a condition that would not
increase the impurity of the body of water into which
it flows; that is, so that the impurities retained shall
bear the same relation in quality and quantity to the
water in which they are carried, as the impurities of the
stream bear to its water. Or again, it may be necessary
to remove all visible impurities and discoloration, while
dissolved putrescible matters, of an amount that would

not cause a nuisance, may flow into a body of water from which no drinking supply is taken.

These different ends may be attained by processes varying in kind or in degree, the selection to be made according to convenience and cost.

In the great majority of cases it will probably be deemed best, especially with a view to possible closer exactions in the future, to proceed at once with such measures as will lead to complete purification, or, when recourse must, for any reason, be had to chemical precipitation, to the most complete development of that method of treatment. The tendency of popular sentiment is toward an increase of cleanliness and comfort, and toward a stricter regard for sanitary conditions, so that the chances are that the demand for a complete purifying of our sewage will increase. This tendency is certainly not likely to diminish, and it will be wise to take this into consideration in establishing the standard of purity to be reached.

This question, as to the degree of purification that is demanded, must be determined at the outset. If it must be complete, then the choice of method lies between irrigation and filtration, for no other process thus far devised will secure this. If something less than completeness is decided upon, then the other methods indicated are also to be considered; and they are variously available, according to the degree of completeness required.

D

CHAPTER IV.

THE TREATMENT OF SEWAGE BEFORE FINAL DELIVERY.

As has already been said, the best results and the easiest avoidance of offensiveness will attend the delivery of the sewage at the outlet in a fresh condition,— before decomposition has set in. This condition is secured by the use of such means of collection and transportation as will prevent all halting by the way, all foul matters being washed completely and constantly forward from the moment of their first rejection to their final discharge. This will involve a constant movement of not far from one and a half miles per hour. With much less velocity there will be a liability to form deposits. To prevent these, even with a rate of fall giving a good velocity, the sewers must be protected against the entrance of obstructing rubbish not susceptible of transportation by the usual dry-weather flow,— at least during the hour when it is at its maximum,— or by special means of flushing provided. A failure to secure this quick and complete removal will result in greater or less foulness, according to its extent and duration.

So far as is now known, all forms of organic decomposition, from the first fermentative softening of the less stable carbo-hydrates to the final nitrification and ultimate chemical change of the ammonia compounds into which the albuminoid matters are first resolved, are effected through the agency of living organisms. These organisms are of many classes and kinds. They act independently or in association, and produce the most various results, according to the circumstances and environment under which they are developed and exist.

Roughly speaking, the destruction of the organic form is the most rapid and complete when the matters undergoing decomposition are most minutely divided and when their particles are most freely exposed to the air, so that each step of the process may take place in the presence of sufficient oxygen, and so that its product may be subjected to the succeeding step under equally favorable conditions. The worst form of destruction, on the other hand, is that where the decomposing matters are in mass, where the free access of oxygen at all points is impossible, and where the work is done by another class of organisms, largely pathogenic, capable of growth without air, under these changed conditions, and producing very different results.

In this latter case, the foulest odors are given off, and compounds of a dangerous character may be produced. In short, we have the worst results of what is popularly known as "putrefaction." In the former case there is a rapid succession of changes, all occurring under the most favorable conditions, without perceptible odor, and reducing the matters acted upon to the condition of plant food, without danger at any stage of the process. The two cases illustrate the best and the worst methods of sewage treatment.

The aim should be to avoid putrefaction, and to secure, as nearly as possible, a complete and rapid resolution with a sufficient supply of oxygen.

Not only should we guard against putrefaction in the work of ultimate disposal; it should be avoided at every stage of removal. Conditions favoring it may exist at various points along the route of travel, all the way from the sink or water-closet to the mouth of the main sewer.

The first point to be considered is the trap by which the outlet of the closet or sink is protected. This should, of course, be large enough to allow free passage for whatever may be discharged, but it must not be so large that it will not be fully cleared of all deposits by the stream that carries the wastes, or by the flush that follows it.

"Pan" water-closets, and others, such as those which are flushed by a spring cock, having a trap four inches in diameter, are often imperfectly washed out at each use, more or less of the solid matter delivered into the trap being liable to retention. Especially where the use of such closets is infrequent, there is a tendency to foul decomposition. This, however, is mitigated when, as is now generally the case, the closet is flushed by an overhead tank, having a quick discharge equal to several times the amount of water held by the trap.

In the case of kitchen and pantry sinks, the condition

is still much less satisfactory. These are rarely flushed in any other way than by pouring into them the dirty water produced in the operations of the household, and the flow of even this amount is retarded by the strainer over the outlet. It is indeed very rarely that the water passes through the strainer more rapidly than it would flow through a trap one inch in diameter, while it is the common custom to use traps having from four to nine times this capacity (two inches or three inches diameter), with waste pipes leading from them which are much too large to be properly flushed by any current they are called on to carry. As a result, the successive accumulation of congealed grease, with an attachment of more or less solid particles, goes on until, by accretion, the channel is restricted to what the usual flow will be able to keep open. In such cases, we have no longer a smooth and direct channel with metallic walls, but a tortuous passage through a mass of putrid grease, often several feet in length, with a constant contribution to the flow of the results of decomposition, and with a constant foul odor in the effluent.

This may seem to be a trifling matter in comparison with the large volume of sewage coming from other sources, but when we remember that in a town of 5000 population there are not fewer than 1000 sinks, we shall see that the amount of organic wastes that they contribute bears no insignificant proportion to wastes coming from other sources, and that it is worth while to deliver it promptly and in fresh condition. The objection to large traps of ordinary form applies equally to the "pot" traps of the Boston plumber. These soon fill with greasy deposits, through which the necessary water-way is kept open, and, as they hold more deposit than other forms, they favor correspondingly increased putrefaction.

The difficulty now being considered is a radical one, and it becomes more and more serious as the volume of retained matters increases, culminating, in the case of the large cesspool, with only an "overflow" connection with the sewer. Such cesspools are very common in many towns. In Newport, probably the majority of houses that are connected with the sewers at all, are connected in this way. The sewer is not used to improve the sanitary condition of the house, only to save the cost of emptying the cesspool. This condition, of course, so far as these houses are concerned, reduces the sewage to its worst and most offensive possible state.

The same objection applies to the use of grease-traps (the larger, the more objectionable) and to all forms of catch-basins, retaining solids and allowing only liquids to pass,— the solids becoming liquid as they putrefy. It is a curious instance of the meeting of extremes that Chicago, which prides itself on its modern civilization, and which gave us such a World's Fair as the world never saw before, permits no house to be connected with the sewer without the intervention of a grease-trap or catch-basin, which reduces all house drainage to its most rotten and stinking condition before it reaches the public sewer.

All this is in direct contravention of the principle
set up in the interest of sanitary improvement and of
decency, and accepted by all sanitarians, that there
must be no foul odor in house drain or sewer, that
putrefaction must not exist anywhere in the whole
system of drainage, that everything discarded in the
house must be carried rapidly to the sewer outlet outside of the town, and that the sewage as delivered must
be in a fresh condition, with its constituents intact and
their full fertilizing value unimpaired.

It is less easy, but it is not impracticable, to secure
as complete absence of putrefaction with the combined as with the separate system. If the inverts of
the sewers are made with a self-cleansing size and
surface, if the grades are carefully and properly regulated, if all street-inlet catch-basins are abolished,
and if the sewers are thoroughly flushed every day with
a volume of water sufficient to purge them of all gutter
rubbish, they will be practically free from objection on
this score.

Many towns comtemplating the introduction of modern sewerage works, have already a greater or less
length of public or private sewers and sewage drains,
often of antiquated and faulty construction, which they
are anxious to utilize in connection with their proposed
improvements. These sewers and drains are rarely fit
for incorporation into a modern sewerage system, and
are equally unfit to be retained, on sanitary grounds
alone. When they are retained, they are pretty sure
to have a bad effect on the character of the sewage

flowing through the new sewers into which they discharge, and on the quality of the effluent to be disposed
of.   These are usually sewers of deposit, and of a
character which justifies their common designation as
"elongated cesspools."

CHAPTER V.

It is not so much the purpose of this book to discuss the reasons for and against "natural disposal," — along with the general drainage of the district, — as to consider the methods of artificial disposal when this is required. At the same time, there are some considerations relating to natural discharge which demand attention in any treatment of the general subject.

Those who best understand its details consider that public economy and advantage are always the most completely served by direct discharge into a sufficient body of water, if no public nuisance is now, or hereafter, to be apprehended, if a source of water-supply is not to be endangered, if the rights of riparian owners are not to be invaded, and if the fishing industry is not to be injured.

One consideration which formerly had so much weight may probably be set aside; that is, the supposed loss of the elements of fertility involved in the pouring of our wastes into the sea. The fact is, that the conditions are exceptional under which we can get back in the crops of a sewage farm more than the cost of mainten-

41

ance, so that we must look for a compensating return mainly to the fact that we ensure the purification of our outflow at the lowest possible net cost. Not only this, but it is now understood that organic matter sent into the river or sea is no more wasted, so far as the economy of the world is concerned, than if it were kept on shore. When our sewage flows off with the drainage, its constituents are, perhaps, quite as likely to come back to us in the form of fish, shell-fish, or seaweed, as they are to come back in the form of crops when it is spread over an irrigation field. In the former case, much of the potential fertility is applied to forms of growth which do not become useful to man, and in the latter, much of it is dissolved out of the soil and carried away (to the sea) by the natural drainage. In intensified filtration, this underground removal is much more complete. In all three of these cases, the universal wasteful prodigality of nature holds its sway, in spite of our efforts to repress it. In either, our success in reaping a return, whether in the fruits of the field or in the fish of the sea, will be in proportion, not to the care with which we adjust the supply of plant food or fish food to the actual demand, but to the abundance from which either form of growth is enabled to take its quota.

There is, therefore, no reason why the authorities of a town on a large river, or on the shore, may not dispose of their sewage in this simplest and cheapest way, provided they can do so without disadvantage to their own people or to others. In determining as to this, the following conditions should be regarded: —

1. The volume of flow of a river, or of the tidal renewal of a harbor, should be sufficient so to dilute the sewage as to prevent its oxidation from becoming offensive.

2. The current should be sufficiently strong to prevent deposits, which might, in time, silt up the channel, or cause a nuisance by their decomposition.

3. The sweep of the current should be such as to prevent the stranding along the shore of unsightly or otherwise objectionable objects contained in the sewage.

4. There should be no suspicion of contamination of water used in the town, or at a distance from it, as a drinking supply, and there should be no considerable fouling of a stream used for watering cows.

5. The fouling of water used in industrial processes, where clean water is required, is actionable. The possibility of this should in certain cases be regarded as a bar to crude discharge into a stream.

Where these conditions can be satisfied, crude discharge, or discharge after very partial cleaning, is often to be recommended. At the same time, we must have an eye to future conditions and requirements, as well as to those of to-day. With a view to these, it is sometimes advisable so to regulate the alignment of the sewers that the possible future need of interception and artificial purification may be met without too much change of existing work.

Due regard should be had to the interests and to the legal rights of riparian owners on the stream below the town, or of littoral proprietors within reach of the tidal movement. In either case, these rights will be more or less affected, according to the volume of the flowing waters. The legal rights of these owners are now tolerably well defined. They are considered in Chapter XV.

The relation of sewage discharge to the fish industry has several important bearings. It has been amply demonstrated that sewage discharged in a putrid condition, or likely to accumulate to such a degree as to putrefy in the body of water, has a very marked influence in destroying fish, or in driving them away; while, if discharged in its fresh condition, and so diluted and so kept in motion as to avoid foul decomposition, it becomes a useful food for them. Huxley, in comparing the productive capacity of the sea and of the land, estimated the product of an acre of arable land or pasture at one ton of grain, or two or three hundredweights of meat or cheese, "while an acre of sea-bottom in the best fishing ground yields a greater weight of fish every week in the year," — the amount of the more important ingredients not being very different in the two products.

Referring to this statement, Sir John Lawes, of Rothampstead asks: "What, then, is the source of this vast amount of manurial ingredients which is taken out of the sea, apparently without return?" He shows that the fertility of a stream for fish, like that of a farm for its produce, is due to its supply of ingredients of substantially the same character. He cites the absence of fish from very pure streams in the Highlands of Scotland, and their abundance in streams which receive the drainage of manured land, of dog kennels, and of other sources of phosphate of lime, etc. He says: "Sewage must therefore largely increase the production of fish, provided that it is sufficiently diluted, and does not interfere with their health." The fish taken by British fishermen alone off the east coast of England and Scotland and the south coast of England, amount to about 600,000 tons per annum, and these contain about 21,000 tons of the three substances, nitrogen, phosphoric acid, and potash,—the removal of which in crops or stock impoverishes our soils.

"So long as fish live upon each other, the sea much resembles a soil covered by a forest untouched by man. Codfish eat herrings, and they in their turn eat smaller fish, but there is no exhaustion of ingredients. . . . There has been a very prevalent impression that the sewage of London has been wasted. The evidence which I have brought forward will, I venture to hope, not only do away with this impression, but will also establish the fact that it has a decided influence on the production of fish."

We need not, therefore, scruple to send sewage to the sea on the score of a waste of fertility. The computation would not be easy to make, but it would probably be shown if it were made that the nitrogen contained in the sewage flowing from all our cities and towns is far less than that in the soluble nitrates escaping in the drainage of fertile lands the country over.

Probably "scruples" in this matter would have very little influence on the conduct of the authorities having the question to decide for a town. The decision would be made on a much more utilitarian basis, and would relate to the simple proposition to dispose of the sewage in such a way as would satisfy present and prospective demands at the least net outlay for construction and maintenance. That is to say, certain sanitary and other requirements being fixed by the conditions of the case, they will be met in the most business-like way, little influenced by the prospective profits of sewage farming, or by the fear of wasting the sources of the nation's wealth.

If we can make a permanently satisfactory disposal of the sewage by pumping it into the sea, into the sea it will go, and ought to go.

CHAPTER VI.

THERE are at least four methods of sewage disposal, from which we may choose according to the conditions governing the case under consideration, and according to our confidence in them. They are: (1) Irrigation; (2) Filtration; (3) Chemical Precipitation; and (4) for very rough work, Sedimentation. Other methods are still the subject of experiment.

1. Irrigation is commonly called "broad irrigation," but it includes also a great variety of furrow, trench, and bed systems, and, in any of its forms, it may, by increasing the volume of the dose, be converted into

2. Filtration, which is only a more copious irrigation.

In both of these methods it is an essential condition that the application be intermittent, so that renewed supplies of oxygen may enter the soil to maintain the requisite oxidizing processes. Without such renewal, the decomposition would take on the pernicious form of putrefaction, described in Chapter IV., which is maintained by organisms capable of living without air, and of which the results are offensive and dangerous. When properly regulated, these methods effect the perfect purification of the sewage, and the first one the best development of its fertilizing value.

47

3. Chemical precipitation comprises all processes by which the natural tendency to the deposition of solids in the sewage when at rest, is increased by producing an artificial precipitate or coagulum in the liquid, which entangles suspended matters, and carries them down. Hundreds of patents have been taken out in England for various combinations of added elements, but those chiefly relied on are lime, sulphate of alumina, and sulphate of iron. Under the best circumstances all suspended matters are removed from the supernatent fluid, leaving it clear, but still containing matters in solution, which are readily putrescible, which produce renewed discoloration during putrefaction, and which have considerable manurial value. Though the bacteria existing in the sewage may be mainly carried down some remain, or are received from the air, and their multiplication in the fertile liquid is rapid. The removal of organic impurities may reach from 50 to 66 per cent.[1] The method is attended with the accumulation of a mass of sludge, which requires a special treatment, and is often offensive.

"The whole art of treating sewage chemically, as it is termed, is to precipitate it while fresh; i.e. before bacterial invasion has so far advanced as to set up active fermentation. When this is once set up, the results are very disappointing."[2]

4. Sedimentation is a process of unaided deposition.

[1] Mass. Report, 1890, p. 791.

[2] Report of Lord Bramwell's Commission on Sewage Discharge (1884), p. 229.

The heavier solids fall to the bottom, but grease and similar lighter matters, which in chemical precipitation are involved in the coagulum and are carried down with it, float at the surface in the form of scum. This is not fairly to be called a method of disposal, save that it serves to withhold the grosser parts of the sewage, which would tend to form deposits and surface pollution in a stream or harbor, in which the finer suspended matters might be tolerated. It is sometimes used for the preliminary purging of sewage, preparatory to irrigation or filtration, but even here it is not to be commended, because a rife putrefaction is its inevitable accompaniment, and its effluent is always foul. Its resultant sludge is much more offensive in manipulation than is that of the chemical method.

Among the somewhat fanciful processes that have been devised, is the "Amines" method, which was described by R. Godfrey, C.E., at the Congress of the Sanitary Institute of Great Britain, at Worcester, in 1889. He said: —

"It is claimed by the inventor of the Amines process — and I am bound to confess with very great reason — that he does effectually destroy the bacteria, and so remove the possibility of any fermentation arising.

"The new reagent is produced by the action of lime on certain organic bases belonging to the group of 'Amines,' or ammonia compounds. When these organic bases are acted upon by lime, a very soluble gas is evolved, which spreads rapidly through every part of

E

the liquid, and is held in solution therein with great
tenacity. This gaseous reagent has been found to be
antagonistic to the existence and propagation of every
species of bacteria occurring in sewage and other similar
waters, for it utterly extirpates them in a remarkably
short space of time.

"The effluent from such water after treatment by this
process is actually sterilized; it shows no living micro-
organisms whatever, even under the most powerful
microscope, and its sterility is further confirmed by the
latest and most severe test known to modern science;
viz. inoculation on nutrient gelatine and plate cultiva·
tion. . . .

"The Amines, from which the process is named, exist
in many substances in nature. And herring brine is
one only of the many sources from which they can be
obtained. They are used either pure or in the form of
Amine salts, or in one of the numerous substances
containing them. But at present the brine is the
cheapest and most readily procurable form in which it
can be obtained. And the reagent formed by its
admixture with 'milk of lime,' and which the inventor
has named 'Aminol,' is a powerful disinfectant, and
imparts a sea-breezy odor to the works in contrast to
the usual foetid effluvia. . . .

"The sludge from this process is of a brownish
yellow color, and lacks the shiny appearance of ordinary
sludge, and from its being permeated with the reagent
(Aminol), it may be left exposed to the sun and wind
without any fear of offensive vapors being given off,

A quantity of about ten tons lying in the open ground at the time of my visit was perfectly inodorous."

The discussion which followed the reading of Mr. Godfrey's paper did not indicate great confidence in the value of this method. Special stress was laid on the fact that to sterilize sewage is precisely what we do not want to accomplish. This is at best only postponing the decomposition, which must, in any case, ultimately ensue, and it is not an advantage, as was claimed, that when the effluent passes into a river it destroys its bacteria also, for the bacteria of river water are busy in effecting the oxidation of its impurities.

Latterly a new process of sterilization by the action of chlorine compounds, produced by the electrolysis of salt (as in sea-water) has attracted some attention in New York under the name of the Woolf process, and in France under the name of the Hermite process. The sterilizing agents thus produced have practically the same chemical constituents as "bleaching powder," and similar results would be obtained by the use of the latter. But the apparent novelty of the process, the popular fascination of electrical methods, and the name "Electrozone" given to these agents, have appealed to the fancy of the public, and results are hoped for beyond the expectation of those who understand the true conditions and difficulties of purification on a large scale.

The fatal objection to "Electrozone," "Aminol," and all other disinfecting and sterilizing compounds, is that, in so far as they are effective, they are pernicious.

What we have to do is to infect our sewage as quickly and as completely as possible, and to make it as fertile as possible for the growth of bacteria. In other words, we must not "pickle" it; we must favor its quick decomposition. Decomposition is inevitable, and the earlier it is completed, the better.

As an illustration: If it were practicable, by the use of electrozone, to destroy every bacterium in the affluents of Croton Lake, a vast amount of food for bacteria would reach that body of water, and processes of oxidation which would have been active and effective in the minor streams would be transferred to the aqueduct and the water mains in the city, where they would be most alarming.

CHAPTER VII.

THE DEGREE OF PURIFICATION REQUIRED.

It is not always necessary to make sewage absolutely pure. General considerations as to this have already been set forth. The degree of purity demanded in any given case will be first of all determined, due regard being had to possible future conditions. Less than this should be regarded as inadmissible; more than this it may or may not be well to seek.

There are cases, as at New York, and at the towns on the lower Mississippi, where there is no sort of objection to the discharge of crude sewage. This is admissible whenever the volume and current of the receiving water ensures a sufficient dilution for oxidation, and sufficient movement to prevent deposits. This does not, of course, apply to bodies of water from which a supply for domestic use is to be taken at any point within fouling reach of the outlet.

There are other cases where it is necessary only to remove the obvious impurities, and such floating matters, like vegetable parings, etc., as would suggest a sewage origin, where real purification is of no consequence, so long as the source of the impurity is not made conspicuous.

53

It is sometimes required only that the sewage, in addition to the withholding of its coarser objects, should be made as clean in appearance as the water of the stream into which it is to be discharged. In other cases, it needs merely to be clarified, which is a very different thing from being purified.

In an increasing proportion of cases, where the ultimate delivery is into a stream from which drinking-water is taken, purification must be as complete as possible. As a matter of prudence, the same condition had better be assumed with regard to streams which are at all likely, within a few years, to be used for water-supply.

In determining the degree of purification to be adopted in any given case, a number of considerations will arise, many of which will relate to local circumstances. From a purely business point of view, the decision may be made largely on financial grounds.

Let us suppose that the natural outlet is into a stream passing through a farming district, and near a town. which may, some day, but not for a number of years. need to use it for a water-supply. The farmers complain of the fouling of the water-supply, especially during summer, when the flow is small. Raw sewage can no longer be sent into the stream, but a moderate amount of cleansing will meet all early requirements. Even this will call for an outlay of say $25,000, and . yearly cost for interest of $1250. To buy land and construct works for complete purification will cost $100,000, involving an annual interest charge of $5000.

Let it be assumed that the cost of maintenance will be the same in both cases, after crediting a certain return in the form of agricultural produce. The difference of interest annually will be $3750. If this interest is compounded at 4 per cent, it will amount to $25,000 in about six years, so that the town can afford, at the end of that time, to sink the whole partial plant (which may or may not be necessary) and make a fresh start, with its complete and permanent works. Then, too, it may be found at that time that the new construction can still be postponed, possibly for a long time. It is also possible that before the new works will be needed some new system of disposal will be devised, which will effect a permanent saving in cost and in maintenance; for this is really a new art, and much is sure to be learned about it as time goes on. These considerations point to the wisdom of tentative methods, and of the postponement of the construction of works for permanent future use,—to say nothing of the uncertainty as to the growth of population.

On the other hand, no engineer likes to build works now which will have to be abandoned, or greatly changed, within a few years, especially as it is by no means certain that his own plans for the change will be followed. He is much more likely to be blamed for the inadequacy of his original recommendations than to be praised for his prudence; and, however clearly he may set forth his reasons for the decision, the public may forget all about them by the time the question comes up of voting $100,000 for the extension. The cost of the abandoned works may also be regarded as so much waste, in spite of the saving of the interest money they have effected.

Then, too, the committee having control of the work may, now that public opinion has been worked up to the point of doing something, be anxious to secure, once for all, what will suffice for permanent uses.

The decision, therefore, as to the degree of purification for which provision is now to be made for a town under such conditions, is beset with no slight difficulty. Still, the ambition of the committee, and the personal interests of the engineer, should not weigh too strongly against unquestionable financial advantages. Cases may arise where temporary works, costing $25,000, may ensure a sufficiently good purpose for so long a time that the interest on the difference of $75,000 will accumulate more than the $100,000 that the permanent works may then cost (about twenty-two years), and a cheaper method may then be available. In all such cases, the committee and the engineer should formulate a permanent project, and put themselves very clearly on record as recommending the temporary one only for sufficient reasons given.

There are many cases, as at Savannah and Mobile. where it is quite likely that artificial purification will never be needed at all, yet where it would not be prudent to disregard the possibility that some degree of clarification may be demanded at a future time, and adequate provision should be made for a proper method. which may be adopted when the occasion requires.

CHAPTER VIII.

THE PREPARATION OF SEWAGE FOR TREATMENT.

SEWAGE, as it flows from the outlet of a system of properly constructed and properly flushed and ventilated sewers, is, as has been said, dirty water, containing some solids which have not yet been broken up by the movement of the current, some fibrous matters, and some rubbish. If the sewers carry surface water, as well as house drainage, the proportion of these will be considerably increased in the dry-weather flow, and there will be an addition of sand and other detritus from the streets.

If the sewers are not thoroughly flushed, if they retain deposits, and if the house-drainage system includes traps and pipes of deposit, or, as in Chicago, grease-traps or settling-basins, there will be a condition of putrefaction. To quote again from Lord Bramwell's Commission, "When this is once set up, the results are very disappointing." Proper preparation for treatment in such cases involves a reformation of the conditions leading to deposit and decomposition,— and no element of the preparation is more important.

The coarser constituents, or such of them at least as will not be broken up in the earlier stages of disposal, should be removed by screening. To remove them by sedimentation in tanks, as is not unusual, leads to putrefaction, and is not admissible in good work. Horizontal screens, where the conditions admit of their use, are better than vertical ones, as they are more easily cleared of their accumulations and are more complete in their action.

After screening, the sewage will still contain a good deal of suspended matter, fibrous and other, which constitutes sludge in the case of Precipitation, and which is an embarrassment in Irrigation, and still more in Filtration. It should be removed as completely as possible in the work of preparation. These constituents of raw sewage have a tendency to attach themselves to the surfaces of any material over or through which they flow. At Wayne, Penn., advantage was taken of this tendency by passing the sewage, when first discharged, over an area of considerable size, which was covered with broken stone (macadam) eight or ten inches deep. This retains nearly all fibrous and gummy matters, and little passes on which would tend to clog the surface of the ground. Nothing passes which is not completely destroyed by exposure to the air in the interval between applications.

Probably something more effective and practical than this method may be devised. For example, two or more alternate shallow channels of considerable width may be prepared, through which the sewage will flow in a shallow stream. These may be furnished with sheets of galvanized iron, having vertical wires, or spikes, higher than the depth of the flow, and placed near together, say one inch apart each way. These will collect the coarser substances, and such of them as are not susceptible of comminution by the current will be retained. These tooth-screens should continue for a considerable distance, according to the volume of the flow. The channels should be narrow enough for the sections of the screens to be of a manageable size, so that they may be removed for cleaning.

Naturally, the screen at the head of the line will accumulate the most, but after its teeth become foul, the lower ones will form their accumulations in succession. The channels thus provided must be amply long, so that the lower screens will still be effective at the end of the periodic flow. It may be practicable to hook the screens together and to set them on a travelling way, so that, as they fill, they may, by suitable mechanical appliances, be drawn up towards the head of the channel and be removed, fresh ones being hooked on at the lower end of the line. On removal, the screens may be laid out to dry, and there will be finally little else than a fibrous felt to be removed from them. Experience with the sludge forked out from the stone beds at Wayne showed that what would be so removed would not be offensive. If it should be found practicable to remove the screens from a travelling way by sections, as indicated, only one channel would be needed. It is a question whether it would be better to use a single channel and remove the screens, or to have two or more channels for alternate use, raking out the accumulations after they have been weathered for a sufficient time. This is suggested as a possible method for removing an element of sewage which gives more or less trouble in final treatment, though much more in precipitation than in irrigation. It would call for another element of construction, but it would be simplicity itself as compared with the method of treating the sludge in precipitation works, and it would remove everything which it is at all desirable to remove in preparing sewage for final purification by land. It would also be good preliminary treatment for sewage that is to be precipitated. Some such device would greatly simplify processes of partial purification, preparatory to discharge into certain streams and harbors.

As a matter of fact, such preparatory treatment opens a field for invention and improvement that is well worth exploiting. The greatest difficulty in sewage disposal — putrefaction aside — is connected with the treatment of elements that may be removed, especially from fresh sewage, as it first flows from the sewer. The embarrassments due to street dirt and to putrefaction under old systems of sewerage have been, or have been supposed to be, a bar to the easy and inoffensive withdrawal of these elements; but, with the fresh outflow of a good system of separate sewerage, so arranged that the disposal works shall not be swamped during rains, there will be no difficulty whatever in developing the foregoing suggestions into an entirely practical method for bringing sewage into a condition that will greatly simplify and perfect its treatment by any process. Indeed, the stone areas at Wayne perform a very satisfactory service.

CHAPTER IX.

THE THEORY OF DECOMPOSITION, AND THE MASSA-CHUSETTS INVESTIGATIONS.

JUST wnat becomes of waste organic matter after it leaves the outlet of a sewer it is important to understand. It undergoes various changes, according to the conditions to which it is subjected, and these affect the rapidity and directness with which it is resolved into its elements, as well as the degree of inoffensiveness attending the various stages of the process.

These conditions and their results have been made the subject of especial investigation by the Massachusetts State Board of Health, at Lawrence, Mass., beginning in 1888, and still continued (1894).[1]

These systematic studies have added greatly to our knowledge, and have given us certain clear indications as to the requirements of practical disposal. They have, in fact, so determined the theory on which success-

[1] Those more immediately connected with these experiments were Hiram F. Mills, C.E., a member of the Board, in general charge; Professor T. M. Drown, chemist, and Professor William T. Sedgwick, biologist, of the Massachusetts Institute of Technology, having general charge of the investigations in their several departments. Mr. Allen Hazen was chemist at the station, and Dr. E. K. Dunham, H. L. Grant, and G. R. Tucker, bacteriologists.

ful practice must be based, as to constitute the most important step thus far taken, and they have made it possible to set forth clearly the natural processes involved, and to indicate certain principles which must be regarded in the artificial disposal of sewage.

In so far as they are not consumed for the mainte-nance of animal life, the organic constituents of sewage are returned to the condition required to make them available for the support of vegetable life.

The final reduction is effected by oxidation. This has long been known, but it was, until recently, sup-posed that there was a direct chemical union of oxygen, as it exists in the atmosphere, with the oxidizable elements of dead organic matter. The theory long prevailed that charcoal especially, and other porous solids, as well, had the power of withdrawing oxygen from the air and condensing it in their pores, and that it was this excessive supply of the destructive gas that gave charcoal, etc., their great destructive power.

The following is from Liebig's Letters on Chemistry (1843): —

"Some gases are absorbed and condensed within the pores of the charcoal, into a space several hundred times smaller than they before occupied; and there is now no doubt they there become fluid, or assume a solid state. . . .

"In this manner, every porous body — rocks, stones, the clods of the fields, etc. — imbibe air, and therefore oxygen; the smallest solid molecule is thus surrounded by its own atmosphere of condensed oxygen, and if in

their vicinity other bodies exist, which have an affinity
for oxygen, a combination is effected. . . . .

"But the most remarkable and interesting case of this
kind of action is the imbibition of oxygen by metallic
platinum. This metal, when massive, is of a lustrous
white color, but it may be brought, by separating it
from its solutions, into so finely divided a state, that
its particles no longer reflect light, and it forms a powder
as black as soot. In this condition it absorbs eight
hundred times its volume of oxygen gas, and this
oxygen must be contained within it in a state of con-
densation very like that of fluid water. . . .

"Thus combinations which oxygen cannot enter into,
decompositions which it cannot effect while in the state
of gas, take place with the greatest facility in the pores
of platinum containing condensed oxygen."

Liebig's theory was that organic matter in a state of
decomposition communicated a corresponding internal
disturbance to other organic matter in contact with it,
their chemical affinity acting under special conditions.
He says:[1] —

"These changes evidently differ from the class of
common decompositions effected by chemical affinity;
they are chemical actions, conversions, or decompositions, excited by contact with bodies already in the
same condition, in which the elements, in consequence
of the disturbance, arrange themselves anew, according
to their affinities."

[1] Agricultural Chemistry, Third Edition, 1843, revised by Gregory,
1847.

He ascribed the action of yeast on sugar to the fact that its smallest particles are in a state of activity, and that they throw the particles in immediate contact with them into the same state, and he supported the theory by analogy.

"But if yeast be a body which excites fermentation by being itself in a state of decomposition, all other matters in the same condition should have a similar action upon sugar; and this is in reality the case. Muscle, urine, isinglass, osmazone, albumen, cheese, gliadin, gluten, legumin, and blood, when in a state of putrefaction, all have the power of producing the putrefaction or fermentation of a solution of sugar."

The theory having already been advanced that microscopic organisms found in decomposing matters might be the cause of decomposition, Liebig says: —

"It has been supposed that this view is opposed to the theory detailed in the preceding pages, which described contact as the cause of a peculiar activity or power. . . . In the theory of fermentation alluded to, it was not asserted that the yeast or ferment could effect the decomposition of sugar at appreciable distances. In this respect, therefore, the two theories are not opposed to each other. They deviate, however, in this, that the one theory considers yeast as a body, the smallest particles of which are in a state of motion and transposition, and that, by virtue of this state, the particles of sugar in contact with it are thrown into the same state of change, while the other theory asserts that the

particles of yeast are little fungi, which are developed
from germs or seeds falling. into the fermenting liquid
from the air, and that in this they grow at the expense
of the substances containing nitrogen, which are thus
converted into, and separated as, fungi. The particles
of sugar in contact with the fungi are supposed to be
converted into carbonic acid and alcohol, which, in
other words, signifies that the act of vegetation effects
a disturbance in the chemical attractions of the elements
of the sugar, in consequence of which they arrange
themselves into new compounds. . . . It is certain
that sponges and fungi, growing in places from which
light is quite excluded, follow laws of nutrition different from those governing green plants, and it cannot be
doubted that their nourishment is derived from putrefying bodies, or from the products of their putrefaction,
which pass directly into this kind of plants, and obtain
an organized form by the vital powers residing within
them. During their growth they constantly emit carbonic acid, increasing in weight at the same time, while
all other plants, under similar circumstances, would
decrease in weight. Hence it is possible, and indeed
probable, that fungi may have the power of growing in
fermenting and putrefying substances, in as far as the
products arising from the putrefaction are adapted for
their nourishment. When a quantity of fungi are
exposed to the temperature of boiling water, their
vitality and power of germinating. become completely
destroyed. If they be now kept at a proper temperature, an evolution of gas proceeds in the mass thus

F

treated; they pass over into putrefaction, and, if air be admitted, into decay, and at last nothing remains except their inorganic elements. The putrefaction in this case cannot be viewed as the act of the formation of organic beings, but as the act of the passage of their elements into inorganic compounds."

Count Rumford had ascribed the green color of water, to which organic matter had been added, to animalcules, "in so much that the green color seemed to be caused by them." Liebig, half a century later, examined water from a trough in his garden, which was "colored strongly green by different kinds of infusoria," and obtained free oxygen from it. He says:—

"Without venturing upon any opinion as to the mode of nutrition of these animals, it is quite certain that water containing living infusoria becomes a source of oxygen gas when exposed to the action of light. It is also certain that, as soon as these animals can be detected in the water, the latter ceases to act injuriously to plants or animals; for it is impossible to assume that pure oxygen gas can be evolved from water containing any decaying or putrefying matters, for these possess the property of combining with oxygen. Now it is obvious, if we add to such water any animal or vegetable matter in a state of decay, that this, being in contact with oxygen, will resolve itself into the ultimate products of oxidation in a much shorter time than if infusoria were not present.

"Thus we recognize in these animals, or perhaps only in certain classes of them, by means of the oxygen which, in some way, as yet incomprehensible, accompanies their appearance, a most wise and wonderful provision for removing from water the substances hurtful to the higher classes of animals, and for substituting, in their stead, the food of plants (carbonic acid), and the oxygen gas essential for the respiration of animals. They cannot be viewed as the causes of putrefaction, or of the generation of products injurious to animal and vegetable life, but they make their appearance in order to accelerate the conversion of putrefying organic matter into its ultimate products."

The Massachusetts Board of Health, in its Nineteenth Annual Report (1888), refers to the results of previous experience on the subject, whose investigation it was then about undertaking. These may be summarized as follows: —

Purification was formerly supposed to be due to the oxidizing effect of the air, but Schloessing, in France, and Frankland and Warington, in England, found that with this there must be the effect of the active presence of organisms to produce nitrification.

Schloessing found that when sewage was passed through baked sand and marbles, no purification was produced at first, but that later the effluent was quite clear and free from organic matter. He found that this purification was arrested by the presence of chloroform in the sand, and that it began again when the chloroform was washed out. This confirmed his conclusion that purification requires the co-operation of organic life. "He concluded that the sewage intro-

duced the nitrifying elements, and that, as their purifying action ceased when treated with chloroform, and
began again after the chloroform was washed out, he
concluded these elements were living organisms."[1]

Warington, believing that nitrification is due to
living organisms, sought to determine their distribution in the soil, concluding that they are chiefly confined to the surface soil.

Frankland concluded that purification is a process of oxidation, producing carbonic acid and nitric
acid, and that a continual aeration of the soil is necessary.

"The conclusion was reached that, in the loam,
nitrifying organisms existed, as shown by Warington,
and nitrification set in at once, while they did not exist
in the sand and gravel, but were introduced by the
sewage, and were retained in passing over the particles
of the sand, where they multiplied till they were in sufficient number to effect the nitrification of the sewage.
. . . The amounts reported as applied to various
filter-beds in England and on the Continent, are from
36,000 to 90,000 gallons per acre per day; but the
analyses of the effluent, when given, are not so satisfactory as those obtained in the laboratories.   From these
results, it appears that filter-beds, if of proper materials,
can purify ten or twelve times as much sewage per acre
as can be applied to any farm lands in irrigation.   It
is upon the basis of these results that we must enter
upon our experiments, to determine the amount of

[1] Massachusetts Board of Health, 1888, p. 39.

sewage we can, in this climate, purify, with such material as is deposited in our valleys."[1]

In their report of the chemical work at the Lawrence experiment station, Dr. Drown and Mr. Hazen say:[2] —

"The organic matters which give sewage its distinctive character are seldom present in the sewage of cities, with abundant water-supply, to the amount of 0.1 of 1 per cent. Yet it is this small amount of organic matter which, by reason of the putrefactive changes which it is capable of undergoing, makes the sewage repellant to the senses, and gives it, either directly or indirectly, its power of producing disease."

Their statement as to the effect of oxidation[3] may be thus summarized: The carbon of the organic matter is first oxidized, leaving the nitrogen as ammonia, which further oxidation converts into water and nitric acid. Both processes may go on at once in the same mass, but this is the order of change with regard to any particle of matter. In sewage, the carbon is partially oxidized, and a proportionate share of nitrogen is converted into ammonia. When sewage flows into a body of water, the nitrogen of the ammonia may, with the oxygen it contains, develop nitrites and nitrates. In porous earth, a more complete exposure to oxygen favors complete oxidation. The process of decay in organic matter is intricate, and involves the formation of many inter-

---

[1] Massachusetts Board of Health, 1888, pp. 40 and 41.

[2] Purification of Sewage and Water, Mass. State Board of Health, 1890, p. 707.

[3] *Ib.* pp. 708 and 709.

mediate products. It is dependent on the life of micro-organisms. Though the supply of oxygen may be unlimited, such oxidation does not go on in nature without these organisms. The chemical action of direct oxidizing agents affects only the carbon, hydrogen, and sulphur of organic compounds, — not nitrogen, ammonia being generally formed. The oxidation of nitrogen requires the presence and vital activity of bacteria. Four forms of nitrogen compounds are recognized: organic nitrogen (albuminoid ammonia), ammonia, nitrous acid, and nitric acid, "this order being that of progressive change from organized to mineral matter."

The fact that the transformation of nitrogen from its organic condition (albuminoid ammonia), through its subsequent changes to that of nitric acid, in nitrates, is effected under conditions favorable to the activity of micro-organisms, in the presence of oxygen, is well shown in the case of a filter-tank at Lawrence, which was filled for a depth of five feet with coarse gravel, none of the stones of which were less than three-quarters of an inch in diameter, and none more than one and a quarter inches. These stones had been washed clean. Sewage was applied daily, in such doses as would cover the stones with a moving film, without occupying the air spaces between them. Concerning this experiment, Mr. Mills says:[1] —

"The experiments with gravel-stones give us the best illustration of the essential character of intermittent

[1] Purification of Sewage and Water, Mass. State Board of Health, 1890, p. 578.

filtration of sewage. In these, without straining the sewage sufficiently to remove even the coarser suspended particles, the slow movement of the liquid in thin films over the surface of the stones, with air in contact, caused to be removed for some months, 97 per cent of the organic nitrogenous matter, a large part of which was in solution, as well as 99 per cent of the bacteria, which were of course in suspension, and enabled these organic matters to be oxidized or burned, so that there remained in the effluent but 3 per cent of the decomposable organic matter of the sewage, the remainder being converted into harmless mineral matter.

"The mechanical separation of any part of the sewage by straining through sand is but an incident, which, under some conditions, favorably modifies the result; but the essential conditions are very slow motion of very thin films of liquid, over the surface of particles having spaces between them sufficient to allow air to be continually in contact with the films of liquid.

"With these conditions, it is essential that certain bacteria should be present to aid in the process of nitrification. These, we have found, come in the sewage at all times of the year, and the conditions just mentioned appear to be most favorable for their efficient action, and at the same time most destructive to them and to all kinds of bacteria that are in the sewage."

This filter, and one filled with gravel-stones of the size of beans, were used mainly to demonstrate, in a very marked way, the fact that the purifying of sewage by filtration is not a process of straining, but of bacterial oxidation. The continued investigations were made with filters containing sand,—ranging from coarse mortar sand to sand as fine as dust,—peat, fine earth, and garden soil.

The sewage experiments at Lawrence were begun in January, 1888. It was already known that if sewage is continually applied to sand in as large quantities as can be transmitted, the sand being constantly saturated, some of the impurities will be strained out and remain in the sand, which will become more and more foul, until, finally, what flows out at the bottom will be as impure as the original sewage. Under this continuous saturation no oxidation can take place, because no air is admitted. Mr. Mills thus describes what takes place when the sewage is applied intermittently:[1] —

"If, however, we change the conditions, and apply to-day, to the surface of a body of open sand, an inch in depth of sewage, and to-morrow apply another inch in depth, and on following days a like quantity, and, after a time, watch what is taking place, we shall find that the sewage, which covered the whole surface an inch deep yesterday, shortly settled below the surface, the bottom particles going down about nine inches, and the top particles remaining just below the surface. In this nine inches, about two-thirds of the space is occupied by sand, one-ninth of the space is water, and about one-quarter is air. The sewage is suspended here in extremely thin layers, covering the particles of sand

[1] Purification of Sewage and Water, Mass. State Board of Health, 1890, pp. 6 and 7.

and stretching between some of the nearest particles, and intimately mingled with more than twice its volume of air.

"Upon covering the surface with sewage to-day, the sewage of yesterday, and most of the air which is associated with it, are pushed down, with more or less mixture, by the incoming sewage, to the nine inches next below; the same quantity of water is pushed down through each of the lower layers of sand, and the same amount goes out at the bottom. Fresh air is brought into the upper nine inches with the incoming sewage, and, if this open sand be five feet in depth to underdrains, the sewage applied on any day will, in very thin laminæ, be slowly moving for a week over particles of sand intermingled with twice its volume of air.

"These conditions are found to be favorable for processes to go on within the sand, which have an effect similar to that of burning up the organic matter, and leaving only a mineral residue in the otherwise nearly pure water. This method of filtration is known as 'intermittent filtration,' in distinction from the first-described 'continuous filtration.'"

It is then shown that there is an important difference between the dry oxidation of actual burning, and the wet oxidation of these laminæ of liquid. In the latter case, but not in the former, nitric acid is formed, which combines with the most available base, forming nitrates of lime, potash, soda, etc. This is called the nitrification of the impurities of sewage, and it marks their final reduction to plant food.

The rapidity and completeness of the result vary with the quality (mainly as to porosity) of the filtering medium, but the principle and the action are the same. Very fine sand or other soil "may remain saturated throughout its whole depth, and give space for air only near the surface, after standing long enough for the liquid there to be evaporated."

It is said[1] that, before these investigations, "very little was definitely known anywhere of the conditions most favorable for the purification of sewage by any given material. Whether it would purify at all, and, if at all, how much sewage could be applied, and what periods of intermission were necessary for obtaining a desired quality of effluent, and whether, and under what conditions, disease-producing germs could be removed, were questions which, at the beginning of these investigations, could be determined only by the resulting purification obtained by experiment. . . .

"The experiment station at Lawrence has been arranged, and the experiments have been conducted, for the purpose of determining the fundamental principles of filtration not previously established, and to learn what can practically be accomplished by filters made of some of the widely varying materials found in suitable localities for filtration areas, that there may be deduced from these results, together with the quality and physical characteristics of the materials used, the probable efficiency of other materials to be found throughout the State.

[1] Purification of Sewage and Water, Mass. State Board of Health, 1890, pp. 8 and 9.

"About four thousand chemical analyses, of the sewage applied to the tanks and of the filtered effluent, have been made in twenty-two months, the results of which are given in accompanying tables."

Corresponding examinations were also made as to the bacteria in the sewage and in the effluent.

"The purifying ability of a filter is indicated, not only by the smallness of the quantity of ammonia in the effluent, but by the greatness of the amount of nitrates. The nearer the nitrogen of the nitrates in the effluent is to the whole amount of nitrogen in the sewage, the more completely has the nitrogenous organic matter of the sewage been destroyed, and its objectionable part been used to form part of an unobjectionable mineral matter." [1]

It would not be possible in this brief space to give such an account of even the most important details of these experiments, and of the resulting deductions, as is desirable. They are fully set forth in the published reports, to which the interested reader is referred.

It may be said, in a general way, that they fully establish the efficiency of filters, when properly constructed and controlled, to purify large quantities of sewage continuously, making it even more pure than many accepted drinking-waters. The account of the detailed investigations is followed by "A General Review of Results," [2] in which the more important features are set forth.

[1] Purification of Sewage and Water, Mass. State Board of Health, 1890, p. 13.

[2] *Ib.* pp. 577 to 600.

It is shown that a sand filter does not effect nitrification when first used. It is necessary for it to accumulate a suitable colony of bacteria, and time is required for this. Furthermore, the colony adjusts itself to the work it has to do. If, then, the amount of sewage is suddenly increased, and is continued at the larger amount, the nitrification will at first be incomplete, but the bacteria will soon multiply, and purification will again become satisfactory, often amounting to the destruction of $99\frac{1}{2}$ per cent of the nitrogenous matters in the sewage, and all but a fraction of 1 per cent of the bacteria. Of the bacteria escaping, it is believed that none are the bacilli of specific diseases, but this is not absolutely stated.

Nitrification is affected by the season and by temperature. It is the most active in the growing months of May and June,—even more so than in the hotter months of July and August. With this exception, the amount of nitrification varies with the amount of the sum of the ammonias in the sewage; "so that, in the winter months of 1888–89, while the nitrates of the effluent were lower than at other times, we find that the sum of the ammonias in the sewage was also lower, and that nitrification at that time was quite as complete as in the previous months."

Even before nitrification begins in a new filter, as with one which was first brought into use " in the cold months of a very cold winter, there was an important step in purification going on. This was the conversion of albuminoid ammonia to free ammonia; or, to state

the case more definitely, it was the burning up of a part
of the organic matter by the combination of oxygen with
some of the carbon, producing carbonic acid, and leaving the nitrogen and hydrogen that were combined with
this carbon to form ammonia, and thus reducing the
amount of combined nitrogen, which in our analyses
appears as albuminoid ammonia. This is as complete
a destruction of organic matter, as far as it goes, as if
the free ammonia were again oxidized, forming nitric
acid or nitrates; but this process seldom, if ever, carries
the destruction of the organic impurities of sewage to
such an extent that the resulting liquid contains so
little impurity as when nitrification takes place. We
find, further, that this process of reducing the albuminoid ammonia is not so destructive to bacteria as the
more complete process of nitrification. It is, however,
a process of purification, and the conditions of intermittent filtration are those most favorable to this step
in purification."

The purifying capacity of some of the materials was
shown to be very great. A filter of coarse sand, which
had been long in use, gave a good result when filtering
at the rate of 117,000 gallons per acre per day (equal
to nearly 2000 population per acre) for nearly three
months, "after which the quantity was increased, and
averaged for five months 177,000 gallons per acre per
day (equal to nearly 3000 population per acre). The
purification was less complete for the first month after
the change, but in the second and third months it was
more complete than with the quantity given above.

The fourth and fifth months, however, gave less satisfactory results,—showing that the filter was becoming overburdened; the surface became much clogged with organic matter, and the sum of ammonias of the effluent increased to 2.7 per cent of those in the sewage, but the bacteria in the effluent decreased to 0.1 of 1 per cent of the number in the sewage. This filter was evidently doing more than it could continue to do indefinitely. The other filters, filtering quantities decreasing with their perviousness from 60,000 gallons per acre per day to 9000 gallons, indicated that they would continue giving as good results indefinitely. In all cases, except that of No. 1, they gave an effluent containing about 0.5 of 1 per cent of the nitrogenous organic matter of the sewage, as shown by the sum of the ammonias, and from 0.08 to 0.001 of 1 per cent of the number of bacteria in the sewage. It is probable that the three less pervious materials allowed no bacteria to be brought down from the sewage, but that the numbers of one or two or three in 100,000 of the number in the sewage grew in the underdrains."

One of the experiments showed the great difference between intermittent and continuous filtration. The filter acting intermittently removed 99.2 per cent of the sum of the ammonias in the sewage. When kept continuously saturated, with the consequent exclusion of air, nitrification ceased, and the sum of the ammonias in the effluent increased steadily, until they even exceeded those in the sewage,—some of those stored in the filter at an earlier stage then escaping. The filter was then

drained and used intermittently. The nitrates soon
rose, until they represented 50 per cent more than the
sum of the ammonias in the sewage, the stored impurities being nitrified. After three months these had all
been burned out, and the effluent had become purer
than the average of the public drinking-waters of the
State.

During the continuous filtration, there was no nitrification to kill the bacteria, "but they were unable to
survive the long passage of three weeks through the
sand without oxygen," their number being reduced
from over 50,000 to less than 100. At the beginning
of intermittent filtration, when there is no nitrification,
but when some of the carbonaceous matter is being
consumed, the bacteria of the effluent average 1 per cent
of those in the sewage, 99 per cent being destroyed;
"but when nitrification begins, the number surviving
the passage suddenly decreases to only 0.08 of 1 per
cent, and a still further decrease to about 0.03 of 1 per
cent when nitrification becomes complete."

Concerning the safety of the effluent as drinking-
water, Mr. Mills says: [1] —

"We have found that the sum of ammonias, which
have been taken to indicate the amount of nitrogenous
organic matter, has been reduced to about 0.5 of 1
per cent of those in the sewage, and is less than the
sum of ammonias of most of the public drinking-water
supplies of the State.

[1] Purification of Sewage and Water, Mass. State Board of Health,
1890, p. 597.

"The chlorine and nitrates are higher than in the public drinking-waters. They indicate in these effluents, as their excess above the normal does in the drinking-waters, that the water which contains them came from sewage; but, in the absence of the ammonias, they indicate that, though the water came from sewage, the organic impurities have been destroyed, and these are merely mineral constituents which remain after that destruction. They are principally common salt and saltpetre, which, in the quantities found in any of the effluents, are regarded as entirely harmless."

Professor William T. Sedgwick, of the Massachusetts Institute of Technology, in general charge of the biological work of the Board, gives the following as the biological aspects of the theory of intermittent filtration: [1] —

"The simplest theory of the working of any filter is that its action is mechanical. Filters which are mere strainers are familiar, and the word 'filter' has come to mean ordinarily a more or less perfect strainer. This primitive idea, however, does not apply to filters such as we are dealing with in this report. A field of sandy soil may, it is true, be a very effective strainer, but, if worked intermittently, it is much more than this. A mere strainer soon chokes, and must be cleaned, but an intermittent filter does not choke, and is self-cleaning. This is a phenomenon which can actually be witnessed. When sewage began to be applied to the several tanks

---

[1] Purification of Sewage and Water, Mass. State Board of Health, 1890, pp. 859 and 860.

outside the station, even the most intelligent of the
workmen predicted that these would quickly choke, and
become a nuisance; but, after two years of actual operation, nothing more remarkable or objectionable could be
seen upon them than upon other fertile land. This
simple ocular demonstration is confirmed by the results
of analysis, and the mechanical theory is readily dis-
proved by a comparison of the chemical composition of
the affluent with that of the effluent. In the life-history
of an intermittent filter, there may be a period at the
outset when there is but little if anything more than
a mechanical purification; but, under the best conditions, there speedily begins a change of the profoundest
significance. The dissolved organic matters no longer
pass out as they came in; the suspended matters for
the most part cease to accumulate, and both appear in
the effluent under other forms. Obviously, mechanical
processes alone could not effect such a change, and,
besides, these changes may occur under conditions which
exclude entirely the purely mechanical hypothesis."

The conditions here referred to are well illustrated
by the experiment with large gravel-stones, already
detailed, and by another, "in which the filtering material was entirely of gravel-stones as large as beans.
The sand had not only been screened out, but all of the
stones had been washed, so that no sand adhered to
them, before they were put into the tank. They formed
a bed five feet in depth, and for nine months sewage
pumped directly from the city sewer was applied nine
times a day, for six days in the week, in quantity

equivalent to 81,400 gallons per acre per day. The quality of the effluent varied somewhat; but during the last two months, June and July, after the above quantity had been applied daily for more than seven months, . . . 98.6 per cent of the organic matter of the sewage, shown by the sum of ammonias, was removed by being burned and converted into nitrates, and more than 99 per cent of the bacteria that were in the sewage were killed. We must regard this as a remarkably good result, with an effluent averaging 70,000 gallons an acre for every day in the year [representing a population of over 1100 per acre].

"The foregoing results were so satisfactory that the quantity was increased by applying the same amount hourly, for fourteen hours, instead of for nine hours. The quantity applied was then the equivalent of 126,600 gallons per acre per day [2100 population per acre] for six days in the week. This quantity was continued for three months, until Oct. 24, 1890, with very little change from the result previously obtained. . . .

"We still find 98.5 per cent of the organic matter of the sewage is removed, and its nitrogen is in the mineral form of nitrates; and more than 99.6 per cent of the bacteria are killed. This result was so satisfactory that the quantity was still further increased in November.

"These results show more definitely than any others the essential character of intermittent filtration. We see that it is not a straining process. By the application of small quantities of sewage over the whole surface of the tank each hour, each stone in the tank was kept covered with a thin film of liquid, very slowly moving from stone to stone, from the top toward the bottom, and continually in contact with air in the spaces between the stones. The liquid, starting at the top as sewage, reached the bottom within twenty-four hours, with the organic matter nearly all burned out. The removal of this organic matter is in no sense a mechanical one of holding back material between the stones, for they are as clean as they were a year ago; but it is a chemical change, aided by bacteria, by which the organic substances are burned, forming products of mineral matter, which pass off daily in the purified liquid.

"The liquid flowing out at the bottom is a clear, bright water, comparing favorably, in every respect that can be shown by chemical or biological examination, with water from some of the wells on the streets of our cities, that are used for refreshing draughts by the public during the summer."[1]

Such coarse gravel could not be used in practical work unless overlaid by a material which would check the entrance of sewage and cause it to descend slowly and uniformly over the whole area. Without this provision, it would flow *en masse* through the first part of the field that it reached.

In the Board's Report for 1891, Mr. Allen Hazen, the chemist of the experiment station, makes the follow-

[1] Purification of Sewage and Water, Mass. State Board of Health, 1890, pp. 670, 671, and 672.

ing very lucid summary of the controlling conditions of the purification of sewage by filtration:[1] —

"The purification of sewage by intermittent filtration depends upon oxygen and time; all other conditions are secondary. Temperature has only a minor influence; the organisms necessary for purification are sure to establish themselves in a filter before it has been long in use. Imperfect purification, for any considerable period, can invariably be traced either to a lack of oxygen in the pores of the filter, or to the sewage passing so quickly through that there is not sufficient time for the oxidation processes to take place. Any treatment which keeps all particles of sewage distributed over the surface of sand particles, in contact with an excess of air, for a sufficient time, is sure to give a well-oxidized effluent, and the power of any material to purify sewage depends almost entirely upon its ability to hold the sewage in contact with air. It must hold both air and sewage in sufficient amounts. Both of these qualities depend upon the physical characteristics of the material. The ability of a sand to purify sewage, and also the treatment required for the best results, bear a very close relation to its mechanical composition."

In applying the teachings of this investigation to surface irrigation, with an inconsiderable amount of filtration, account is to be taken of the fact that conditions requisite for nitrification may be created in the surface soil by evaporation. It is also to be considered that, in proportion as evaporation takes the place of

[1] Massachusetts Report, 1891, p. 428.

filtration, in like proportion, at least, will the carrying down of bacteria into the underdrainage be reduced.

As to the question whether or not the piling up of the scrapings of the surface of a sewage-filter bed, containing large amounts of organic matter, will create a nuisance, Mr. Hazen says:[1] —

"To this we can answer, no. The stored matters are the most stable portions of the sewage; they have resisted strong oxidizing action, and are incapable of rapid or objectionable decomposition. The matters which would have caused trouble had they been stored are just the ones which have been oxidized. The material should be so placed that a change of air in its pores will be possible, and no offence need be anticipated."

This accords fully with the experience at Wayne, Penn., where the accumulated rubbish forked out of the broken-stone straining beds was placed in a large heap, with no other sand or earth than had come from a separate system of sewerage, causing no offence whatever.

It was found that peat, as used in a filter at Lawrence, was almost useless, as the quantity of sewage that would pass through it was too small to render filtration practicable. The flow through peat five feet deep, and with a surface area of one two-hundredth of an acre, averaged, for twenty-two months, only eight and a half gallons per day, or 1700 gallons per acre per day.

---

[1] Massachusetts Report, 1891, p. 454.

This is so at variance with natural conditions as to call for further consideration. An ordinary peat swamp, drained only by open ditches at wide intervals, will absorb the water of the heaviest rain almost as it falls, and can be walked over soon after the rain ceases. A rainfall of two inches is equal to over 54,000 gallons per acre. A portion of the sewage-disposal field at Freehold, N.J., is of nearly pure peat. Before drainage it was an impassable morass for some days after a heavy rain. It was underdrained, at a depth of four feet, by three-inch tiles, laid at intervals of twenty-five feet, and it dried out rapidly. Soon after this improvement, a rain of over two inches fell on it in one day. The following morning carts were driven over it without difficulty. It has since shown no inability to absorb sewage in large doses.

It is probable that the availability of peat as an absorbing and filtering medium in its natural condition, depends on some peculiarity of its internal structure, which is destroyed when it is removed and worked over. At Lawrence, "this material was at first completely saturated with water, by cutting it up fine and sprinkling it into water standing in a tank." The slow deposition of fine material suspended in water would lead to an arrangement of its parts very different from that existing when a fibrous mass has slowly accumulated by the decay of vegetation.

Very unsatisfactory results were also obtained in experiments with garden soil and finer earths. These too, were different from what we obtain under natural conditions, and the difference may have been due to similar changes of interior construction.

It is regretted that the account of the Massachusetts investigations can only be summarized in this inadequate way. All that is possible in the space here available is to set forth their simplest and most useful teachings, and so to stimulate curiosity as to incite a careful study of the extended official reports.

CHAPTER X.

It may be assumed, after ample concurrent testimony, that the only way in which sewage can be really purified on a large scale — purified to the condition of safe and attractive drinking-water — is by bringing it into contact with the particles of some porous material, such as earth, sand, or gravel, under conditions which will expose it to an adequate supply of oxygen. This is done, in practice, by flowing it over a suitable area of suitable land, and by allowing such periods of rest between the flowings as will permit the water to be so far removed as to admit air throughout the mass of material which the organic matter of the sewage has penetrated.

The processes of purification which sewage undergoes in such material were succinctly set forth in the last chapter. Attention will now be given to the way in which these processes may best be availed of in actual work.

According as the volume applied is small or great, the method is called Irrigation or Filtration. The dividing line between them, as has already been said, is somewhat vague, especially when sandy and gravelly soils are considered. Perhaps the safest distinction would be to say that so long as the degree of flooding does not interfere with proper agriculture, the process may be called Irrigation, and when this limit is exceeded, Filtration; but even this definition is not exact.

With the more porous soils, crops will grow well with very large doses of sewage, while those of closer texture, with less capacity for absorption, are practically not available for filtration, although they may receive, utilize, and purify a large irrigating flow.

It is generally good practice in irrigation disposal to provide three areas of land, to be flooded on successive days,— one day of use with two days of rest. With this arrangement, even stiff clay soils will, in time, come to do well with an application equal to one inch or more in depth over the whole surface during the twenty-four hours of use. If of lighter texture, they will take correspondingly more.

For a rough-and-ready calculation, it will be sufficiently exact to say that 1 inch over an acre is 25,000 gallons, and that, with an allowance of about 60 gallons per person per day, this will equal the sewage of a population of about 400. By the same token, 2 inches would give 50,000 gallons and 800 population, and 3 inches would give 75,000 gallons and 1200 population. Dividing these amounts by 3, as each area is used only one-third of the time, we have 8333 gallons and 133 population with 1 inch, 16,666 gallons and 266 population with 2 inches, and 25,000 gallons and 400 population with 3 inches. In this latter case, the average daily application would be equal to a rainfall of 1 inch in 24 hours, and would require the absorption and evaporation of that amount in that time.

This question will be considered hereafter. It is the purpose here to discuss only possibilities. Attention is called to the fact that any agricultural soil will improve in its capacity to absorb and purify sewage as its drainage improves with time, and as its texture becomes modified by cultivation, and by the increase and deepening of the growth of roots, and of the channels formed by the decay of roots. To this extent we may expect future results in excess of what is at first possible, and the improvement will usually be very considerable.

The development of a good texture, and the maintaining of an open surface, will be facilitated by the removal from the sewage of its coarse sludge-forming elements; the finer ones will be taken care of by natural processes attending exposure during periods of rest. This branch of the subject is discussed in Chapter VIII.

After such preliminary treatment applied to the sewage as it reaches the farm, it is in the best condition for irrigation. In fact, so far as distribution and absorption are concerned, it may be used like so much ordinary irrigation water, save that it is even more important here to secure even distribution, in order to take advantage of the oxidizing capacity of the whole field.

In this discussion, agricultural returns will be left

entirely out of the consideration, reference for this branch of the subject being made to Chapter XI. At the same time, it is to be remembered that the evaporating power of the leaves of vegetation, and the penetration of the soil by roots, are useful factors of successful irrigation. Whatever is done, therefore, and however copious the flooding may be, every detail should be made as favorable to growth as possible, and methods of application should be adjusted to the manner of cultivation that is to be followed.

It cannot be too often repeated, that the utmost care must be taken to secure uniformity of flow, the avoidance of ponding, the regularity of alternation and cleanliness at every point. These not only tend to inoffensiveness and to fertility, but they affect the completeness of purification as well.

An acre of land contains 43,560 square feet. To fit a field for sewage disposal, it is desirable that it should be drained, naturally or artificially, to a depth of at least 5 feet (6 feet would be still better). Almost any soil so drained will contain at least 30 per cent of voids filled with air and available for the admission of water; that is to say, the voids of an acre of land, drained to a depth of 5 feet, will admit about 480,000 gallons of water.

For proper purification, it is desirable that the water should occupy about six days in its descent through the 5 feet of depth. The daily dose required for this would be about 80,000 gallons per acre. If the area is divided into three sections, for alternate use, the daily dose for each would be 240,000 gallons per acre. On an average, the upper 6 inches of cultivated land has not far from 50 per cent of voids, equal to (say) 80,000 gallons per acre. The rate of descent through the ground, controlled by the slow movement near the bottom of the filter, being 10 inches per day, the descent of the two days of rest would vacate the voids of 6 inches of surface and of 14 inches of subsoil, and the descent during the day of application would equal another 10 inches. In the 24 inches of subsoil, the capacity of the voids (at 30 per cent) would equal about 195,000 gallons, making the total space available for absorption sufficient for 275,000 gallons. It is assumed that near the surface the descent would be much faster than 10 inches per day, 20 inches being already vacated, and a greater absorptive capacity being due to the existence of root channels, etc. This would then be the *theoretical* capacity for such a soil with such a rate of filtration. Such a calculation can be only theoretical. Many porous soils would allow a much more rapid descent of sewage, while more compact soils would greatly retard descent. The case supposed implies a soil rather freely absorptive for the upper 3 feet or more, and sufficiently compact at the bottom to prevent a descent at a faster rate than 10 inches per day.

In the above calculation no account is taken of evaporation, which is constant and, during the growing season, very large. Marié-Davy found in his experiments with the filtration tanks of Paris that, of 24,000 cubic meters of sewage applied in six months, only 1600 cubic meters reached the underdrains, all the rest being evaporated from the surface of the ground and by the leaves of the crops.[1] It would seem, therefore, that it may be assumed that land which is at all suited in its original composition for sewage irrigation, and is properly prepared for its work, will transmit or evaporate during the growing season 80,000 gallons per acre per day, if the sewage is so applied to it as to reach all parts of its surface with sufficient completeness. At 60 gallons per head, this would be the product of a population of 1333.

This capacity would not be reached by heavy soils, but it would be more and more nearly approached as continued use improved the absorptive power and the drainage of the material forming the subsoil. On the other hand, it might well be exceeded with many kinds of soil, of which the voids amount to more than 30 per cent. It is to be understood that we are still considering theoretical capacity.

In order to estimate properly the duty to be imposed

[1] The Sewerage Committee of the British Association reported in 1871 on experiments conducted by them at Breton's Farm (Romford) "under such conditions as would ensure the collection of the whole of the effluent water" to determine the amount of evaporation. The observation covered more than a year (399 days). The average amount of sewage pumped on to the farm per day was 1182 tons. The average rainfall was 695 tons, making an average daily application of water equal to 1877 tons. "The average amount of effluent water discharged was found to be 513½ tons per diem." That is to say: In a moist climate, in an examination which covered a whole year and the wet month of March besides, of all the water reaching the farm as sewage and as rain, 72.65 per cent was evaporated from the surface of the ground and from the leaves of vegetation.

on an irrigation field, it is necessary to determine the volume of sewage to be dealt with and the uniformity or irregularity of its daily flow. With great variations in this regard, close calculation as to area is not possible. The advantage of uniformity is so great, and the cost of all disposal is so considerable, that engineers are fast coming to the conviction that, wherever irrigation, filtration, or chemical precipitation is to be resorted to, the separate system of sewerage should be adopted, — excluding all storm-water from sewers of which the flow is to be treated. If this is done, and if even the most casual effort is made to prevent waste of water, 60 gallons per head per day — being more than twice the average *use* of any mixed population — will be an ample allowance for the amount of water that will find its way to the sewers. It is therefore assumed that, where sewers are to be built in conjunction with any form of artificial disposal, the separate system will be adopted. If the combined system is used, another basis of calculation must be taken.

When we study practical work abroad, we see that by far the most numerous instances of irrigation treatment of which records are available are those where the sewers are on the combined system. Even what are called "separate sewers" in England are meant to carry a large volume of rain-water, usually all that falls on the back roofs and paved courts, and often much more; so that, during storms, the volume of discharge is several times multiplied. The application, therefore, of foreign experience to results to be expected where the entire exclusion of rain-water prevails, as in this country, must be made with care, and even then we shall be able to deduce only probable inferences, rather than trustworthy indications; but, in formulating practical work, we must give ourselves the benefit of the doubt and avoid too close figures. It is as important to the town as to the engineer that the margin for safety be not lost sight of. The provision of land should always be well beyond what existing experience and an estimate of the volume to be treated may indicate as safe. At the same time, to spend money for an unnecessarily large area is neither good engineering nor good business, unless with a view to profitable farming. In making computations as to the relative cost of the different methods of disposal, it is inexcusable greatly to over-estimate the area needed for irrigation, and to decide from this estimate that irrigation must be discarded and a more costly and less effective process adopted.

There is no "rule" as to the area required. From the point of view of the sewage farmer, the more the better; from the point of view of the engineer, so much as is really needed, and not very much more. Sewage disposal by irrigation is much more widely used in England than elsewhere, and it is natural that we should be influenced by English practice in regulating our own works. The English notion that one acre of land should be furnished for each 100 of the population has been accepted here as a safe precedented limit. This is all right when that area can as well as not be had, but its acceptance has subjected more than one town to a heavy outlay for chemical precipitation, when a correct estimate of the needs of the case would have given it the benefit and economy of irrigation.

England is a rainy land, and farmers are often troubled to ripen and save their crops with only natural conditions to contend with. At such times, even the amount of dry-weather flow of the sewers would be only an added difficulty. The combined system increases it occasionally, and in the wettest weather "by fifty or more times," and the separate system (*vide* Santo Crimp) by eleven times. Under these conditions, the demands of "sewage farming" call for a wide extension of land.

Gray, in his table of statistics, gives the following instances of larger rates than 100 per acre: Bedford, 142; Croydon, 144; Dantzic, 250; Breslau, 400. These are all cases where there is a large increase of volume during rain. At the London Sewage Farm at Barking, according to the calculation of Mr. Morgan, the manager, "the sewage of no less than 335 persons was passed through every acre." [1] At Merthyr-Tydvil, in 1888, with good agricultural results, the 262 acres of irrigation land disposed of the sewage of 100,000 persons, or 382 per acre. At times, for weeks together, the 20 acres of filtration area at Troedyrhiew took care of the sewage of 40,000 population, or 2000 per acre. Eliot C. Clarke, C.E., says that, at the State Penitentiary at Concord, Mass., 80,000 gallons were taken up by three-

[1] Journal of the Royal Agricultural Society, 1871, p. 404.

quarters of an acre of gravelly soil. At 60 gallons per head, this would be at the rate of over 1750 per acre. Dr. Alfred Carpenter, contemplating much storm-water flow, says that there should be not less than one acre for 250 population. The farm at Wrexham, which, with that at Bedford, was awarded the first prize of the city of London, and where storm-water is considerable, has one acre to 166 of population. At Swanwick, 15 acres of clay soil take, in addition to storm-water, the sewage of 4000 persons, or one acre to 266. In connection with the sewage purification works of Paris there are four experimental tanks, each 33 feet by 28 feet, containing about 6 feet in depth of earth taken from the irrigation grounds. These received, for six months, sewage at the rate of 232 persons per acre,— only 6.6 per cent escaping through the drains, which indicated that much more might have been taken. Sewage has been applied at Gennevilliers at the rate of 100,000 cubic meters per annum per hectare, which is at the rate of one acre for 487 of the population.

In most of the instances referred to above, the sewage is spread over the surface of the ground. The use of the "ridge and furrow" system, where it is run into trenches between narrow beds, on which plants are grown, affords an easily controllable method of treatment. It may be suitably so arranged that the channels for sewage, filled to a depth of 1 foot, should hold 100,000 gallons per acre, or, one-third of an acre filled every third day, 33,000 gallons, or the sewage of 550 persons per acre. This is the method adopted at

H

Gennevilliers, where the amount of sewage applied is regulated *solely* by the economic needs of the cultivators. They have no obligation to take, for the sake of purification, more than they require for their crops. Yet, under these conditions, they sometimes use it at the rate of nearly 500 persons per acre, and, on the average of the whole year, at the rate of more than 150 per acre.

To come now from the theoretical to the practical, and to keep well within the demonstrated limit, it is safe to assume that, according to its capacity for absorption, a well-regulated irrigation field will, without precluding the possibility of a considerable return in crops, absorb and purify the sewage of a population of from 250 to 500 per acre; and that, as cropping is put more and more into the background, the population to the acre may be increased until, in the case of porous but fine sandy soils, it reaches from 1000 to 1500 per acre.

If clarification, rather than purification, is the object, then these limits may be considerably raised.

To secure the most satisfactory results, it will be desirable so to adjust the flow over each area, or over each of several sections into which each area may be divided, that it shall be sufficiently copious to reach at each application to the furthest limit of the land, with perhaps a small excess which will sink into the ground in a short time.

In regulating the application, if the flow is copious, as from a town of considerable size, it may be run directly on to the area; but if it is not well in excess of the absorptive capacity of the ground, then it should be accumulated in a self-discharging tank or reservoir, which will hold back the whole flow until it is full, and then deliver it in a large stream.

The first work to be done in preparation for irrigation is so to regulate the surface and the drainage of the land as to fit it for the maximum of service. Unless the ground is underlaid by a porous subsoil, which will ensure the free removal of all water settling through its upper layers without reappearing in the form of "springs" on its lower parts, it should first of all be so thoroughly underdrained as to secure this condition.

If the drains are made deep enough, say 4 feet, to ensure the maintenance of the line of saturation during the latter part of the period of rest at a depth of 3 feet from the surface, this will ensure a condition of absorption and of aeration which will suffice for purifying a large volume of sewage, but it would be better to go considerably deeper if practicable. The drains should be larger than ordinary agricultural work would require, — not less than 3 inches in diameter, and not necessarily larger, unless, in a rather porous subsoil, they are to be placed more than 40 feet apart. While 4-foot drainage may suffice, it is better, if a suitable outlet can be had from that depth, to lay the tiles 5 feet, or even 6 feet, below the surface. Even in soils through which water would at first settle slowly to this depth, the drainage will improve with time, until, even in clay, it will become effective.

The outlet, though at the cost of some decrease of depth, should run perfectly free,— that is, the full diameter of the mouth of the tile should be above the level of the water into which it is discharged,— so that the freedom of its flow may be watched.

To this end, also, it is best, when practicable, not to collect the drains into a main or intercepting tile drain, but to deliver them independently into an open channel. This should not be a mere open ditch, with steep sides liable to cave and to fall, but a channel which may easily be kept clear.

As an illustration of this, reference may be made to the arrangement of the sewage-disposal field at the Women's Prison at Sherborn, Mass. (1879). The field had a slight slope toward one of its boundaries, which had a fall of about 3 inches per 100 feet. The under-drains, at right angles to the contour of the land, or in the line of steepest descent, were 5 feet deep. They discharged into a channel which was 1 foot wide at the bottom, and of which the sides were carefully graded to a uniform slope of $1\frac{1}{2}$ to 1, and were well sodded. The channel was 5 feet deep and 16 feet wide at the top. The water-way was furnished with a 2-inch plank, 16 inches wide, with 2×4 strips, spiked to its outer edges, which they overlapped 2 inches at each side, as shown in the accompanying sketch, Figure 1. The slope of the banks came to the inner edges of the 2×4 strips, and held the whole in position. A cross-section of the whole channel is shown in Figure 2. The woodwork was always completely saturated, and had a constant stream flowing through it. It was thus protected from decay, and remained serviceable until the field was abandoned on the recent construction of a local system of sewers.

The great advantage of this method lies in the fact that each underdrain is thus always open to inspection. It is also satisfactory to be able to watch the promptness

Fig. 1—PLANK BOTTOM

with which sewage applied at the surface increases the drainage flow, and the fluctuations of volume under different doses, and during rains of various amount, as effected by the condition of different parts of the field.

In constructing underdrains, care should be taken to lay the tiles to a true grade, and they should be pro-

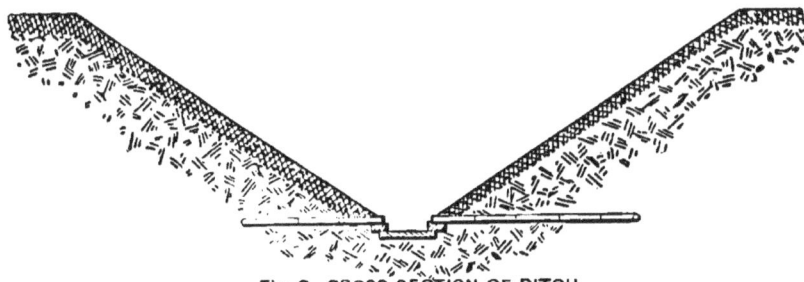

Fig. 2—CROSS SECTION OF DITCH

tected against the entrance of silt by being wrapped twice around with strips of cheap muslin. The middle of the strip is laid under the joint, and each end is folded over the top and tucked under the tile. The muslin will last as long as it is really needed, for, after the ground becomes thoroughly settled and compacted in place, there will be no direct flow from above into the drains. They will receive their water from below, as the general level of the ground water rises into them from the pressure of that which settles down through the "lands" between them.

This settling of the earth over and about the tiles is most important, and too much attention cannot be given to it. Even with the most careful filling and the most thorough ramming, the line of the ditch will be likely to furnish water-ways, and there will thus be a liability to a direct flow of sewage into the drains without full purification, and with danger to the stability of the drains themselves. The safest plan, where water is to be had, is to cover the tiles, to a depth of about a foot, with fine material, well tramped into place, and then to fill the trench with water and shovel the remainder of the backfill into this. This secures a good settlement at once, and prevents the direct flow of sewage into the tile.

The proper regulation of the surface is more or less difficult in proportion as it is more or less important. If we have a uniformly sloping surface, considerably larger than is needed, and covered with growing grass, it will probably be necessary only to lower a few hummocks and to fill a few depressions, removing the sod at these points and replacing it after the grading is finished. This can be done by any skilful laborer who

has a straight eye in his head; but, on the other hand,
we sometimes have to deal with a flat meadow, with
clumps of bushes, depressed cow-tracks, boulders, and
stumps; or it may be a steep hillside, with trees of all
sizes and in all stages of growth, some of them fallen
and some of them rotting on the ground; or a very
irregular surface, with outcropping rock, areas of clay
and areas of gravel, and sloping in different directions.
In such cases, skill, judgment, and experience will be
needed, not so much to secure a good final result, as to
secure it at a reasonable cost.

It is, of course, possible to reduce any surface and
any material to any desired grade, and to fit it for
irrigation. An engineer accustomed to railroad grading will know how to do this, but such work is costly,
and it is often quite unnecessary; the result may be no
better than if the irregular natural conformation is
generally maintained, only the minor inequalities being
corrected and actual "pockets" being filled.

The importance of grading increases with the reduction of area. If there is land far in excess of the needs
of the population, we may take full advantage of its
better portions, and make less use of its less favorable
portions. If, on the other hand, the area is limited,
then it becomes more important to make all parts of it
equally available. It is by a proper application of this
principle that economical good work is to be secured.

If the land is flat, it must be made quite smooth and
uniform; and, if surface irrigation is contemplated, it
must be worked into beds or broad ridges, with draining

furrows between them. This will be most necessary
with a heavy soil. If it is of a sufficiently absorptive
nature, it may be best to arrange it in narrow, flat beds,
separated by parallel ditches into which to discharge
the sewage for lateral absorption. A field that is
sufficiently sloping, however irregular it may be, if
made smooth as to its surface, can have the sewage so
led to all its parts as to work uniformly.

While the preparation of an irrigation surface involves
no difficulties that may not be overcome, there is nothing within the whole range of sanitary engineering in
which so much depends on skill and good judgment.
The work must be done once for all, and it must be so
regulated as to get the best result for the least outlay.

The land being levelled and graded, and being covered with some sort of vegetation, where this is necessary to prevent washing, the next step is to arrange for
the distribution of the sewage in such a way as to secure
the full use — or at least the equal use — of all parts
of the land. As the application must be intermittent,
even the smallest irrigation tract should be divided into
two or three different areas, and as it increases in size,
— especially as it increases in relation to the volume of
sewage to be disposed of, — still further division may
be desirable.

In making such division it is not always sufficient to
lay off the farm or the field in a certain number of tracts
of equal size. The division should be made, not according to area, but according to absorptive capacity. The

sections of heavy soil, in which absorption is slower,
should be larger than the sections of light soil, in
which absorption is rapid. Land that is nearly level,
and over which the flow is slower and accumulates until
it reaches every spot, saturating the surface completely,
will drink in more sewage than steep land, over which
the flow will be rapid, and where there is more tendency
to form rills, where artificial runnels are needed to
counteract this tendency, and where spots which are
higher than their neighbors will get less than their due
share of the supply.

The needs and conveniences of cultivation must also
be kept in view, and in many cases — but by no means
in all — one or more relief areas should be provided,
which will bear heavy flooding on occasions when the
requirements of cropping make it desirable that sewage
be kept from tracts whose regular turn it may be to be
irrigated. All of these conditions are to be taken into
intelligent account and satisfied as completely as is
practicable.

The division and arrangement being decided on, the
next step is to provide for the proper preliminary treatment of the sewage, and for its proper distribution over
the various areas. Reference is here made to Chapter
VIII.

The particular method of preparatory treatment to be
adopted in any case will depend on the lay of the land,
and on the manner and volume of delivery at the field.
If broken-stone areas are used, they should be so
placed, and should be of such size and form, as will

cause a reasonably uniform flow of the sewage over and through them, so that when it leaves them it shall have been freed of most of its fibrous and more adhesive contents. The current will wash out from among the stones some of the results of the decomposition of earlier deposits, and may become darkened in color thereby, but this will be no detriment; the matter thus added to it will no longer be in such condition as to impede absorption, and the discoloring material will soon be taken up by the ground over which the sewage flows.

In the case of land of any considerable inclination, unless quite regular and smooth in its general conformation, there will be a tendency to irregular flow, which would lead to uneven distribution. To counteract this tendency, it is necessary to arrest the stream at intervals,—more frequent as the slope becomes steeper,—and to collect it for a fresh start. It has been usual to accomplish this by means of horizontal gutters or ditches, following the contour of the land. These ditches fill with sewage and overflow uniformly along the whole length. The objection to their use lies in the fact that, after the flow ceases, they retain sewage for evaporation or slow filtration, and so lead to more or less putrefaction and consequent foul odor. This may be avoided by substituting for such depressions banks of porous material, laid on the proper contour lines on the undisturbed surface of the ground, or on pavement.

Such banks may be made of coarse gravel, or, better

still, of broken stone that has been passed over a half-inch screen to remove its finer parts. With a little attention to its upper side, adding or removing slight obstructive matters, such a bank may be made equally effective along its whole length. A considerable portion of the sewage will flow through it, but more or less of it, according to the volume of the flow, will pass over its top. The irregularities of porosity and of top level will tend to compensate each other, and the distribution will be more uniform than if the flow were all through a porous mass or all over the top of a bank of earth.

The greatest advantage of the porous barrier is that it allows all of the sewage to pass with such freedom that nothing is retained to putrefy. Under proper management — chiefly the removal of leaves and grass, and the occasional raking of the back of the bank — the land at the upper side of the barrier will be dry within an hour after the flow ceases, and there will be no more odor here than over other parts of the flooded area; that is to say, there will be no odor at all.

These porous barriers are not to be considered as strainers, as performing the function of the beds of stone over which the sewage first passes. When it reaches the first barrier, it will already have lost most of its solids, and it will coat the surfaces of the stones of the barrier so slightly that all traces of it will have passed away before the next application of sewage.

The flow decreases in volume as it passes over the land, a portion of it being absorbed as it goes on; therefore the upper barriers may, with advantage, be coarser, and the lower ones finer, according to the volume to be transmitted. The lower barriers at Wayne, made of locomotive cinders loaded with stone at the top and on the lower slope, are never overflowed; the small amount of sewage reaching them all passes through them. Stone barriers have the further advantage that they require little attention to keep them in shape. They do not wash away like earth, and they are not disturbed by the frost. They are apt to be penetrated by a rather vigorous growth of weeds, which must be cut or pulled from time to time, but this is no disadvantage from the point of view of disposal.

This general equalization of the flow will not be all that is required, except on nearly level and smoothly graded land. Water flows on the line of least resistance, and the slightest depression or obstruction suffices to divert it. It will usually, therefore, become necessary to regulate its movement between the barriers by forming occasional V-shaped leading runnels here and there, often only a few feet long, or to fill depressions with a few spadefuls of earth, or both. This final regulation can best be done while sewage is actually flowing, the movement that it takes indicating the slight amendments to be made.

The main discharge of sewage at the beginning of the field may be led to the stone beds of the several areas either by covered pipes or by open channels. If these latter are used, they should be made of half-pipes of vitrified earthenware, laid to the proper grade, with their joints filled with clay,—not with cement when frost is to be anticipated. They may have outlet branches at fixed points of delivery, or they may be made a little lower on the side toward the field, so as to overflow for a considerable length, the stream being dammed from point to point, as one stretch after another is to be reached. These half-pipes should be bordered on each side with well-laid sod, on a true grade, and provision should be made for complete emptying of the channel soon after the flow ceases.

The combinations that may be formed of various means of straining and distribution are to be regulated according to the special indications of each case. As an illustration, the writer is now arranging plans for the irrigation of a long, narrow disposal field, bordering a stream which must not be fouled. The land will have to be divided into rather small areas, on lines running at right angles to the contours. The length of the field is some 3000 feet, while its average width is not more than one-fifth of this. It is desirable to use as little of the width as will suffice in rough preliminary treatment. It is also necessary to make the overflow uniform along the upper side of each area. The arrangement under consideration is shown by the accompanying illustration, Figure 3.

An open channel conveys the sewage from the screening chamber along the upper side of the field, for such distance as may be necessary for providing sufficient straining beds. As these beds require extension for future work, the channel will be extended. It has side outlets, controlled by stop-gates, discharging on to separate sections of straining area designed for the first rough cleaning of the sewage, after the manner of the broken-stone areas at Wayne.

The arrangement here will be more systematic. These beds will be paved with brick, and each section will have sufficient vacant space to allow the stone to be forked over from time to time, without the inconveniences that attend such forking on bare ground. The paved flooring will allow the stone to drain more freely and to become more freely aerated. It will possibly be found advantageous to have four different sections, each serving for one day's use, but only three of them being in the rotation at any one time,—one always taking a long rest, perhaps of a week, a fortnight, or a month, as the case may be, and becoming perfectly purged of its accumulations.

In the illustration, the first two sections are shown ready for use; the third, in process of being forked over; and the fourth paved, but not yet furnished with stone. These beds are level longitudinally, but have a slight fall laterally, discharging into a channel, provided with side outlets, controlled by stop-gates, leading to the irrigation areas. For the length of the stone beds (and for any future extension of them) this channel should be level, so that the flow of any section may be led to any irrigation outlet; but continuations beyond the straining beds may have a slight fall.

At each discharging point, a broken-stone area, or apron, say ten feet square, should be provided to check

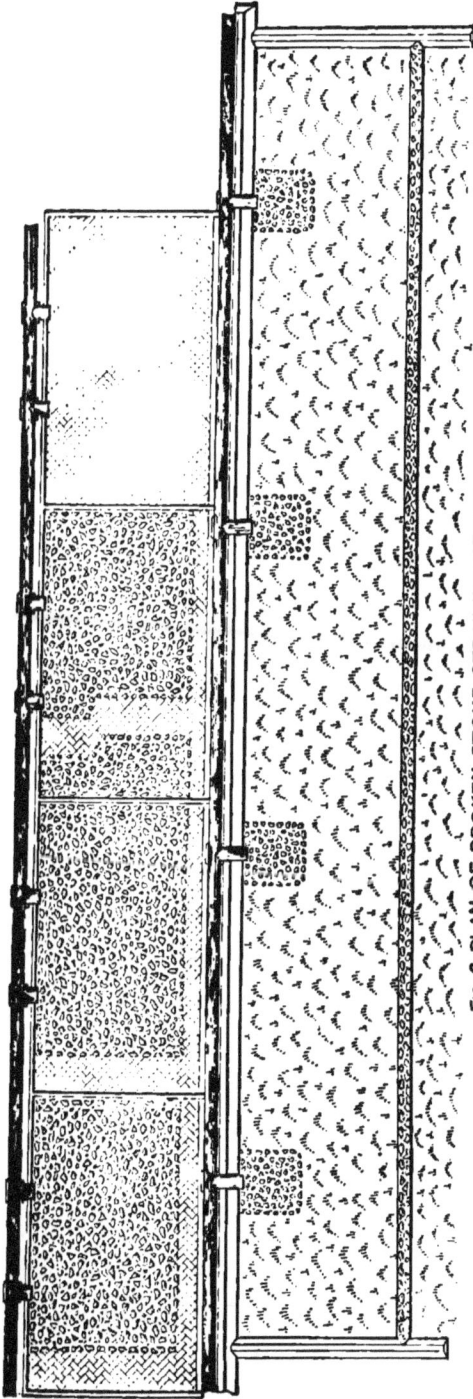

FIG. 3-PLAN OF BROKEN STONE STRAINING AREAS

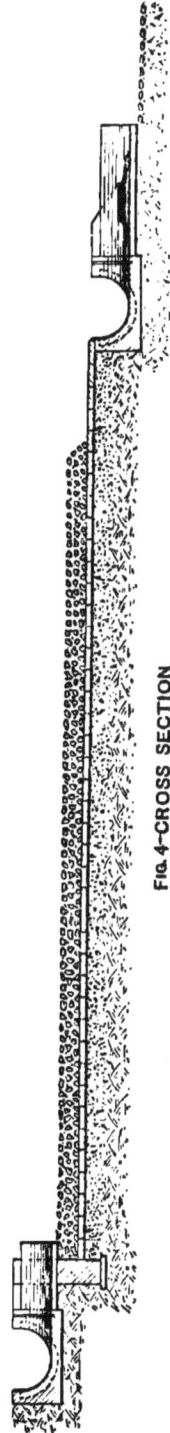

FIG. 4-CROSS SECTION

111

the flow and to prevent washing or ponding. The illustration shows a stone barrier to equalize the flow, and earth embankments separating the sections of the irrigation field.

Figure 4 shows, in cross-section, the two channels and the broken-stone bed, with a part of the apron below the outlet to the field.

By a proper adjustment of the stop-gates, the sewage may be delivered over any one of the screening sections, and discharged over any area of the land at pleasure. The distribution over the land will be regulated, as previously described, by stone barriers or otherwise, as the requirements of the case may suggest.

The arrangements for discharging the sewage over the land in proper condition and with proper uniformity being completed, or, in other words, the work of construction being done, the work of maintenance begins, and it would be desirable that the man who is to have charge of this should have taken an active part in the previous regulation of its details, that he may be familiar with everything connected with the system. The engineer who has directed the work should keep oversight of its practical operation until he has assured himself that all goes smoothly and well.

Dr. Frankland says, in the First Report of the Rivers Pollution Commission, 1870: —

"A field of porous soil, irrigated intermittently, virtually performs an act of respiration, copying on an immense scale the lung action of a breathing animal; for it is alternately receiving and respiring air, and thus dealing as an oxidizing agent with the filthy liquid which is trickling through it."

This method of action should be kept constantly in mind, and the irrigation field should be maintained always in the best condition for its effective performance.

I

CHAPTER XI.

SEWAGE FARMING.

WHILE it is true that, in deciding as to methods of sewage disposal, we must not be swayed by the hope of too much return from well-regulated sewage farming, and while cash income must always be held secondary to purification, yet, at the same time, any help that may be secured from this source towards paying the necessary expenses of the work will be very welcome. We are not, therefore, properly qualified to determine the main question, until we know all that we can learn as to the agricultural considerations which affect it, and which must not be neglected in calculations of cost.

It is hardly fair to say, as is sometimes said, that sewage farming is an art by itself. It is the art of farming with the art of irrigation added. So far as the disposition of the sewage is concerned, beyond what would be required for the best growth, the irrigation is a detriment; but it cannot be avoided, and it must take precedence. The farming must be as good as it can be made in spite of this, but no consideration of cropping must be allowed to interfere with the chief aim of securing complete purification,— or such partial purification as the circumstances of the particular case may demand.

114

If the conditions are such as to make it worth while to give prominence to the agricultural element, there need be no difficulty in securing good management. Farming is not a trade that it is difficult to learn, and the director of the disposal works will, if not a farmer himself, be able to secure an assistant who is. If he is a farmer, there are no details of sewage irrigation and filtration calling for any special knowledge which he may not readily acquire. These details are much less exacting than are those of chemical disposal works, where economy requires a constant testing of the quality, and measuring of the quantity of sewage, and the constant adjustment of the chemicals to the requirement of the moment; where a steam plant and filter presses are to be kept in order and in economical service; yet these scientific processes, where a man specially fitted for such duties is indispensable, are no more a reason for rejecting this method of treating sewage, than is the requirement for a capable manager for an intricate system of water-supply, involving the use of pumping machinery and a complex system of filters, a reason for not providing good water; or than the requirement of a man of first-rate practical capacity as chief engineer of a modern fire plant is a reason for not making the best provision against conflagrations.

In neither of these cases, any more than in sewage farming, will the interests of the town be left at the mercy of an incompetent man.

Let it once be decided that sewage is to be disposed of in any given way, and *let the need for special skill in*

its treatment be appreciated; there will be no difficulty
in finding a competent person to carry on the work in a
satisfactory way.   For example, our many engineering
schools are turning out more graduates than there is a
demand for, and many of them, men of excellent character and of fair capacity, are destined to remain in the
lower ranks of the profession until they drop out and
take to some other occupation for which they are better
fitted.   Some of these men are born and bred farmers.
If a manager were needed for a sewage farm, and if the
selection were to be made from men with fair engineering qualifications, with a practical knowledge of farming, with a capacity for controlling subordinates, and
with good character and habits, there would be no
dearth of applicants to choose from,— men who would
be glad of a permanent place, reasonable pay, an interesting occupation, and a position of independence and
authority.

It ought not to be necessary to say that competent
management is indispensable to success.

If a sewage farm — or any other farm — is to be
entrusted to a man whose chief claim to his position is
that he needs its shelter and its salary, failure is inevitable, as would be, under the same conditions, the
failure of a precipitation plant, water-works, or fire
department.

Even the most competent manager must be made to
learn that it is the first purpose to secure a satisfactory
disposal of the sewage, and only the second to get
money returns from the farm; that he must not l ·

tempted, by the hope of harvesting a promising crop in the best condition, to turn a week's sewage on to land not fitted to receive it. He will learn in time to cut his coat according to his cloth, and to grow the best crops that a proper handling of the sewage will allow. This must be accepted as the best measure of his skill and capacity. The rule should never be lost sight of, by the committee or by the manager, that everything must yield to the proper disposal of the sewage, and that only such farming is to be done as will not interfere with this.

The first important question that presents itself is that of the amount of land to be provided, not only for the present population, but for the growth of the next twelve or fifteen years. To look beyond that period involves a question of interest money, of uncertainty as to the rate of growth, and of possible new invention, which may make it wiser to leave the remoter time to take care of itself. The area required will also depend much on the character of the land and the degree of purification to be secured. In any case, however, if land is plenty and cheap, it may be well, with a view to extensive farming, to secure enough to furnish an acre for even every fifty of the population. Sewage farming becomes a comparatively simple matter with such an abundant provision, and may usually, near a large town, be made a source of some income. If, on the other hand, cheap land is not available, then we must consider the safe requirement for purification, without much regard to agricultural return. It is quite common, as we have repeatedly said, for engineers to accept blindly the not too well defined English notion, that there must be one acre to each one hundred of the population. As a matter of *necessity*, this estimate is not well founded, and it becomes sometimes important to know to what extent it may be disregarded. This subject is fully discussed in Chapter X.

There is really no very well-marked line of distinction between irrigation and filtration, for so much of the sewage applied in irrigation as is not evaporated filters through the ground to the underdrainage, and when sewage is applied for filtration it only irrigates the land more copiously. In a general way, however, it may be said that irrigation is a help to farming, while intensified filtration is a detriment to it, and here the line is drawn somewhere within the limits between benefit and injury. If the amount of land is so restricted that the ripening and harvesting of crops are interfered with by the volume necessary to be taken care of, it is a great help to have a filtration area for occasional use, sufficient to take the surplus flow for a week or more at a time. This will sometimes serve as a regulator, and enable us to maintain purification without sacrificing a crop that is nearly ready for market.

In looking over the literature that has accumulated during the past forty years, with a view to determining the method of cropping best suited for sewage farming, one reaches the conclusion that, as a rule, and with the exception of cereals, what is best suited for cultivation in the ordinary farming of the region is suited for use with irrigation. This is more the case when the amount of sewage applied is small than when it is great. Still the range is large of crops which will bear heavy flooding, under conditions which prevent ponding and too continuous surface saturation. Probably the most important crops, however, are grass, cabbages, mangolds, and celery. These will thrive under a large amount of flooding and yield enormous returns; but if the surface is kept from long saturation, many other plants do remarkably well. The course of cropping best suited to each case can be determined only by actual trial. Much will depend on the character of the soil, on the climate, and on the market. The chief reliance has been in many cases on Italian rye grass, or on this and timothy mixed.

Wherever the land is level enough to be worked into flat beds, with ditches between them,—on the Gennevilliers plan,—no matter what the character of the soil, any crop may be grown with success, but such as will bear the cost of hand harvesting will be best suited for this arrangement of the surface.

Narrower ridges, for single rows of plants, with deep furrows between them for the sewage, are capital for field crops of cabbages, mangolds, etc., but here the sewage must frequently be withheld for a period long enough to allow the furrows to be ploughed out and to kill the weeds,—which grow rampant on a sewage farm. In this method of cultivation, the ridges may follow the contour of the land, but even this arrangement is best suited to tolerably uniform surfaces.

For ground that is irregular in its slope, plain surface flowing is the most available method for applying the sewage, and grass is the most suitable crop, though at Leamington, fine crops of mangolds were grown under these conditions.

On a nearly level farm, it is possible to arrange the surface in "lands." At Breton's farm this was adopted almost universally. The lands were thirty feet wide and as long as possible. They were formed by back furrowing, and were made one foot higher (more or less, according as the soil was light or heavy) in the middle than at the edge of the furrows dividing them. The sewage was led to them along narrow gutters at the centre, and overflowing at both sides. Such an arrangement is somewhat costly at the outset, but it is very convenient for subsequent work, as ploughs and carts can be run from end to end of the lands.

The great growth of grass and of other fodder crops that results from sewage fertilization, is very profitable when the product can be sold green or consumed on the farm, but it is often difficult to make it into hay, unless there is much more surface available than is needed for the mere disposal of the sewage. It is not unlikely that the new method of preserving forage by ensilage may be found very useful in such cases, especially when the production of milk is an important part of the industry. The silo has proved itself most valuable for preserving corn fodder in a green (and half-fermented) state, and there would seem to be no reason for doubting its applicability to the storing of the grass of a sewage farm.

Concerning this, Professor Henry Robinson says in
a paper on "Some Recent Phases of the Sewage Question":[1] —

"If the green crop, after being placed in the silo, be
left freely exposed to the air for a few days, it heats,
and if the temperature be allowed to rise to from 125°
to 150° Fahrenheit, the bacteria are killed, and the
subsequent fermentation which would have been produced by that cause is arrested. The resulting material
is called 'sweet' silage, which remains aromatic, after
having gone through what is called a 'haying fermentation.' If, on the other hand, as soon as the green crop
is put in the chamber, it is subjected to pressure, by
which the air is excluded as far as possible, and if the
temperature does not rise above 100° Fahrenheit, at the
outside, the bacteria live and develop, and the fermentation is greater than before, leading to the formation of
lactic or acetic acids, with a loss of some of the
saccharine matters, which become broken up and form
new combinations, which partly pass away as gases.
The silage thus produced is called 'sour,' to distin-
guish it from that termed 'sweet,' although the difference does not affect the avidity with which cattle eat it,
nor its nourishing results. The fermenting action
which takes place in the silo is partial digestion, such
as would have taken place in the animal's stomach. It
causes a softening of woody fibre and a preservation of
the flesh and fat producing ingredients of the green

[1] Treatment and Utilization of Sewage, Corfield and Parkes, pp.
418 and 419.

crop. The loss of weight produced by fermentation
falls upon those parts of the fodder which have the
least feeding effect, such loss comparing favorably with
preservation of grass, etc., by exposure to the sun."

Any town needing a sewage farm ought, however, to
furnish a demand for green forage, to be carried away
for cows and work-horses, so that here, as in England,
it might be sold standing, to be cut and hauled off by
small users or dealers.

There would seem to be no reason why the osier
willows, now so largely imported, should not be grown
in this country, and they are especially well adapted
for sewage farming. There is hardly a limit to the
amount of flooding that they will bear, and they are
much used in England for relief areas, provided to
receive sewage that must be temporarily withheld from
ripening crops, or during harvesting or rainy periods.

The degree to which the authorities of a town should
engage in the business of a farm is a question to be
settled according to local considerations. All that was
done by the city of Paris was to set the example and to
demonstrate, in its experimental garden and fields, the
value of sewage as a manure, and the value of its water
in irrigation, and to encourage private cultivators to
use it. Such a course may perhaps sometimes be followed here with good success, and land and sewage may
be let out to small tenants, under restrictions which
will not interfere with the proper disposal of the flow.
This arrangement would be especially suitable for truck
farming. Where grass and root crops are chiefly grown,

the farm might furnish stabling and forage to a dairy-man owning his own cows and carrying on the milk business on his own account.

The main point is, and this is important, that the town should retain absolute control of the disposition of the flow of sewage, and should carry on its work of purification untrammelled by any consideration of profit from the working of the farm,—so arranging this part of its operations, of course, as to favor the farming as much as is safely possible.

Whatever is done, due attention must be paid to cleanliness, neatness of appearance, and absence of odor. Whatever is in the least degree objectionable about a sewage farm is *always* due to bad management and neglect.

The possibilities of a large return from sewage culti-vation are great, and it cannot be doubted that they may sometimes be so developed in continued practice as to constitute a source of considerable income.

One of the best, and best-recorded, instances of good sewage farming in England, is that of the farm at Bed-ford, which divided with Lieutenant-Colonel Jones's farm at Wrexham the £100 prize offered by the Man-sion House Committee in connection with the London International Exhibition of the Royal Agricultural Society, for the best-managed sewage farm in England and Wales, utilizing the sewage of not more than 20,000 people.

This farm was established in 1868. It contains 183 acres, 153 acres being irrigated with sewage. It is

light land on a gravel subsoil.   The preparation of the
farm cost £6950.   The sewage is delivered by pumping,— for 123 acres to a height of 13 feet, and for 30
acres to a height of 21 feet.   It is screened at the
pumping station.   The average daily quantity is 1,140,-
000 U.S. gallons.   The distribution is through earth-cut
channels, ploughed or dug out from time to time, as is
required.

A great variety of crops is grown, market-gardening
being combined with farming.   The total produce for
1875, 1876, 1877, and 1878 ranged from £2225 to £2428.
The income and expenditures of the year of examination
(1878) were: —

INCOME.

Valuation of Stock and Plant, 31st December, 1878,	£ 1375	2	0
Sales of Crops,	1884	18	10
Rent of Meadow Land,	119	17	0
	£ 3379	17	10

EXPENDITURE.

Valuation of Stock and Plant, 1st January, 1878,	£ 1237	1	0
Wages,	573	15	7
Manager's Salary,	145	0	0
Horse Corn, Keep, and Straw purchased,	9	14	7
Seeds and Plants,	144	11	9
Auctioneer's Commission and Expenses,	73	18	0
Miscellaneous Bills,	104	12	4
	£ 2288	13	3

This leaves out of account rent, taxes, cost of pumping, and produce consumed on the farm.

The report is well summarized in W. Santo Crimp's
"Sewage Disposal Works," [1] from which the following
statement is extracted: —

[1] Pages 130 to 141.

" *Horses.* — Six strong Shire-bred horses are employed on this farm solely for working the land and for other farm operations. In the summer, the horses are principally fed on rye-grass, and in the winter on beans, oats, and chaff, with a considerable amount of roots, carrots being generally used until Christmas, and mangolds afterwards. The manager reported that this course of feeding keeps the horses particularly healthy. He also reported, in answer to special inquiry, that he never had a greasy legged horse on the farm, and does not consider the horses on the sewage farm more liable to that, or any other disease, than on farms under ordinary cultivation. He further reported that the veterinary surgeon's bill on this sewage farm, during the last four years, averaged about 3s. 6d. per horse.

" *Sanitary Result.* — Eight persons reside on the farm, six of whom are children, and about twenty men and boys are engaged on the farm who do not reside on it. Mr. J. H. Collett, the farm manager, reported: —

" 'No man, either living or working on the farm, or any man living near, has ever suffered from any epidemic disease.

" 'No man ever employed on this farm for a sufficient length of time to have felt the ill-effects of sewage has died up to the present time, neither has there been a death of any resident, young or old.

" 'I can give you no other information on this point, excepting that my men have been particularly healthy, and I have never heard the men employed on this, or the adjoining farms, or any person living near, ever complain of injury or annoyance from our utilization of sewage.' "

The Craigentinny meadows (Edinburgh), including a considerable area of poor sea-sand, have been irrigated, or parts of them at least, for some two hundred years. The amount of sewage applied is enormous, and no regard is paid to purification, only to production.

"Often as much as ten or fifteen thousand tons per acre during the growing season, besides an indefinite quantity during winter. . . . The stream flows on in almost undiminished foulness to the meadows lower down. . . . It is plain that an enormous quantity is applied, much beyond the needs of the largest possible crop of grass. . . . The average [rental] value of the land, irrespective of the sewer-water application, may be taken at £3 per imperial acre, and the average rent of the irrigated land at £30, making a difference of £27; but £2 may be deducted as the cost of management, leaving £25 per acre of clear annual income due to the sewer water. During the past year, the highest price attained was £41 17s. 6d. per acre,[1] and from that down to £19 an acre has been realized. The Italian rye-grass on the same farm has varied in price from £32 an acre for the first year's cuttings, to £25 an acre for the second year's cuttings. . . . The summer's grass of the lower Craigentinny meadows is sold by auction to the Leith and Edinburgh cow-keepers every spring, and the maximum value reached last year

---

[1] That is to say, that the receipt from a certain quarter of an acre was at this high rate.

was £36 15s. per statute acre. The quantity of grass for which such prices are obtained is believed to vary from 50 to 70 tons per acre. And as the means are perfected of distributing the sewage more evenly, and as the subsoil drainage of the land improves, the quantity and price are both increasing year by year. No exhaustion is apparent anywhere. The sewage brings down more than the plants require of every necessary constituent of their food, so that even the poor sea-sand is as fertile as the rest, and the land is getting richer year by year, notwithstanding the enormous crops it yields. Taking the average price of the whole 240 acres to be £24 an acre, we have a total annual produce of £5760 a year extracted by the land and grass from the drainage of 80,000 people."[1]

In the report of the judges of the Royal Agricultural Society, on the sewage-farm competition of 1879,[2] it is stated that, judging from the accounts furnished by the competing farms, "the profit or loss on a sewage farm is almost entirely dependent upon the amount of rent, rates, and taxes which are paid." They conclude that "practically, there appears to be no great value in sewage itself; but that, given an ordinary farm and a sewage farm at the same rent, the sewage farm will hold its own even in a wet and backward year like the past, but in dry periods the sewage farm has a much greater advantage over an ordinary farm. As a mode of effectually disposing of sewage in an innocuous manner,

---

[1] First Report Rivers Pollution Commissioners, vol. i., p. 75.

[2] Journal of the Royal Agricultural Society, vol. xvi., part i., p. 2.

and generally in an economical way, an examination of
the several farms, and of their accounts, shows that the
system pursued is most successful and satisfactory. The
advantage and economy of sewage farming, as a mode
of dealing with sewage, are shown very conclusively in
the case of Birmingham, in which the farming operations show a profit of £1064 18s. 7d. in the year 1878,
while the chemical treatment of the sewage, two-thirds
of which is passed into the streams of the district after
such treatment, cost in the same year £11,987 15s. 3d."

Speaking of sewage farming at Aldershot, Herbert J.
Little said:[1] "Here land of absolutely no agricultural
value by nature may be found producing its crops of
rye-grass up to 50 tons per acre per annum, letting for
£20 or £25. [$97 or $121.25.]"

Naturally the greatest interest has always been taken
in the effect of sewage farming on the health of those
who work and who live on or near the farms. A careful investigation of all the records must result in the
conviction that,— save possibly where the arrangement
or the management is grossly defective, and even here
serious ill effect on health seems doubtful, unless there
is an actual ponding of putrid sewage,— the process has
no detrimental effect of any kind. Careless management is very apt to result in offensive odors, but this
is no argument against the *system*. Where a farm is
converted into a marsh, the usual diseases of marshy
districts may prevail; but marshy conditions have no
place in the proper management of land irrigated with

[1] Journal of the Royal Agricultural Society, 1871, p. 396.

sewage. The British Sewage Commissioners reported concerning their examination of the irrigation fields at Milan, that they find "no evidence whatever of the slightest injurious tendency of irrigation conducted with the waters of the Vettabbia [sewage irrigation], beyond those of other districts around, and where plain water is employed."

The sum of all the evidence makes it clear that diseases which are communicated through the discharges of the bowels, such as cholera, typhoid fever, dysentery, and diarrhœa, are not communicated by sewage irrigation, even among the families living and working on sewage farms.

The most serious evil that has been feared in the past quarter of a century was the communication of entozoic diseases, due to the known vitality of the germs of tapeworm and other entozoa. Dr. Cobbold, the highest authority on this class of infection, expressed, in 1865, great fear that these diseases would be spread by sewage irrigation. Professor Corfield and Dr. Parkes, in their work on the "Treatment and Utilization of Sewage,"[1] after a long discussion of this subject, say: "We see no reason, therefore, to alter our opinion that it has not yet been shown that sewage irrigation has ever increased the amount of entozoic disease in men or cattle, still less that it is likely to do so to a greater extent than any other method of utilizing human excrement."

A governmental Board, called the "Bundstag," where representatives of all the German States discuss impor-

[1] Edition of 1887.

tant measures, considered, in 1886, a proposition that irrigation farms, for the utilization of sewage, should be placed on the list of institutions requiring special permission for their establishment. The Bundestag declined to entertain this proposition. "This means that the ruling authorities in Germany know that sewage irrigation can be carried on without becoming any nuisance, that purification of sewage by sewage irrigation is of immense importance to public health and to the state of the rivers, and therefore they declare it permissible anywhere." [1]

After all these years of experience, it may be stated in the most positive manner that there is no sanitary objection whatever to the system of sewage disposal by agricultural irrigation, and that no nuisance or offence can arise in connection with it, save as a result of gross neglect or mismanagement.

[1] Treatment and Utilization of Sewage, Corfield and Parkes, p. 493.

CHAPTER XII.

THE purification of sewage by filtration is an outgrowth of experiments made by Dr. Frankland and of the recommendations of the Rivers Pollution Commission. It has been found that filtration by an upward movement through a sand filter, which was used for water-supply, would not suffice for sewage, because it excluded air from the filter. Water that is only slightly impure contains considerable free oxygen, while sewage practically contains none. It was also found that air is carried into the filter with a downward flow, and that when the flow is suspended the liquid settles away and is followed by air, which fills the voids thus left in the filtering material, ensuring a supply of oxygen; also, that by a properly regulated intermission, an effective combination of this oxygen with the impurities of the sewage is secured.

This process, as formulated by Dr. Frankland, was called Intermittent Downward Filtration, but as all sewage filtration, as now carried on, is downward and intermittent, the simple name Filtration is used.

As the scientific principles governing filtration are the same as those which govern irrigation, and as the details of the two processes have only technical differences, Chapters IX. and X. are referred to, as setting forth the principles, and indicating in a general way the processes according to which this more intensified method of treatment is to be regulated. This chapter is therefore devoted to such modifications of arrangement and manipulation as the purification of large volumes of sewage on small areas of land requires.

Any general consideration of sewage filtration should take account of the services of the late J. Bailey Denton, C.E., who, in his work at Merthyr Tydfil (Wales), in 1871, gave the first practical demonstration, on a large scale, of the soundness of Dr. Frankland's deductions. Mr. Denton published, in 1881, "Ten Years' Experience in Works of Intermittent Downward Filtration." In this the experience at Merthyr Tydfil is set forth. Concerning the character of land suitable for sewage cleansing, he says: —

"The most suitable soil for both irrigation and filtration is a sandy loam with a small proportion of gritty gravel to quicken percolation. The soil most unsuitable are very dense clays, bog-peat, and very coarse gravels. . . . The superiority of loamy land, properly drained, consists in the affinity for ammonia which its clay constituents possess, and the extreme comminution to which it is reducible by the action of air and water. Under proper treatment, a loamy soil becomes not only more productive of vegetation, and therefore a better purifier of sewage, but it constitutes a better filtering material (mechanical) than either gravel or coarse sand.

Clay soils are not to be recommended for surface irrigation, and can only be used for filtration by an outlay in
draining, earth-burning, and mixing, which sanitary
authorities are indisposed to expend.

"The best soils for intermittent filtration are those of
a free character, closely pulverized with such a proportion of alumina equally dispersed through their bulk
as does not exceed 7 per cent of their constituents.
Sewage will pass but slowly through such description
of soil, and they therefore require quickening by carefully designed underdrainage to overcome their natural
retentiveness. . . .

"The capacity of soils to absorb water is no criterion
whatever of their cleansing capability, whilst their
retentive powers exercise great influence on the rate of
percolation and the quality of the effluent. A coarse
gravelly soil thoroughly drained, for instance, will
absorb and discharge liquid almost as quickly as it
reaches its surface, and will give out an effluent but
imperfectly purified, whereas a loamy soil, having a
sufficient proportion of sand to render it free and to fill
it with close interstitial spaces for aeration, will discharge a satisfactory quantity of purified water by the
underdrains, and maintain a very superior effluent.

"Experience conclusively shows that, while some
soils, even in their natural unmoved condition, will
let sewage pass through them too quickly, others have
retentive powers — I speak of clay and peat soils —
which not only retard in an unfavorable degree the
passage of water through them, but in some degree

injuriously affect the effluent itself, by rendering it
cloudy or discolored, though not chemically objectionable.

"The effect of pouring liquid on soils when charged
by attraction, is to drive out by the fresh liquid that
which is already in possession of their interstitial
spaces; and as these spaces can hardly be said to be
perfectly aerated (though it can only be by the influence of the atmosphere existing in the soil that the
water is driven out), the action is not as perfect as
desirable.   A peaty soil is the most retentive, and at
the same time the most absorbent of soils.   It will hold
water of greater weight than itself, but it readily yields
when in this saturated condition to the gravitating force
of liquids applied to its surface.   Clays of the denser
nature . . . will stubbornly resist its passage through
them.   It is this condition that renders them the most
unsuitable of all soils for filtration, though by burning
and mixing they may be rendered available at a cost
which, though comparatively great, may yet be less than
that of other treatments."

The facts as to Merthyr Tydfil, as Mr. Denton sets
them forth, may be thus summarized: —

The population of 1871 was about 50,000, of which
the sewage equivalent to that of 25,000 was discharged
at Troedyrhiew.   The total dry-weather flow varied
from 860,000 to 1,200,000 [U.S.] gallons per day.   The
larger amount was doubled in wet weather.   It was
proposed to take 375 acres of the valley of the Taff.
Of this, about 75 acres was near Troedyrhiew, 2 miles

from Merthyr, and the remainder 8 or 10 miles further down the valley, to be reached by a very costly outfall sewer. It would take so long to perfect this work that riparian owners suffering from the pollution of the river applied to the courts for relief in the meantime, and Mr. Denton was appointed to effect the remedy.

For temporary relief, he suggested intermittent downward filtration, as propounded by the Rivers Pollution Commissioners, selecting 20 acres of the free soil of the Troedyrhiew farm. This tract was divided into four areas of 5 acres each, and was drained to an average depth of 6 feet. Each 5 acres thus contained 48,400 cubic yards of filtering material. The conduit bringing the sewage to this was also available for the remaining 55 acres.

Mr. Denton says in his pamphlet (1881): —

"For five months the sewage, equivalent to that of 25,000 persons, was put on the 20 acres, and it was so effectually cleansed, that when Dr. Frankland analyzed the effluent on the 20th of October, 1871, the amount of organic nitrogen was found to be 0.012, and the amount of ammonia 0.025, or 1 part in 100,000 parts. In 1872, the same eminent chemist, confirmed by Drs. Benjamin Paul and Russell, found the amount of organic nitrogen to vary from 0.014 to 0.033, and the ammonia from 0.060 to 0.095.

"The filtration areas, though laid out as a temporary expedient, did their work so completely, that it soon became apparent that if the whole of the comparatively small farm of Troedyrhiew (75 acres) had been properly prepared, it would have been ample in itself to have cleansed the whole sewage of Merthyr for the next 30 years without recourse to the 300 acres of distant land.

"In fact, it is certain that, had not the notice of purchase been already served on the landowners, and the works themselves commenced, the local Board of Merthyr would, as soon as the effect of the filtration works was seen, have limited their operations to the Troedyrhiew farm alone."

Mr. Harpur, the local engineer, in his testimony before the Rivers Pollution Commissioners in 1872, said that, when the whole sewage was being poured on to the 20 acres, he noticed no tendency to choke up, and that, if they decided on sacrificing the vegetation, he thought they might cleanse the sewage of the whole town. He had not noticed any sign of the land being overdosed, so far as the cleansing was concerned. In his testimony before the Lower Thames Valley Main Sewerage Inquiry, Mr. Harpur said that the sewage of Merthyr Tydfil, Aberdare, etc., with a combined population of 100,000, was being dealt with on 212 acres, including the 75 acres at Troedyrhiew, part of which was laid out for filtration and part for irrigation. This was equal to a population of 470 per acre.

Mr. Denton says: "By this combination of districts and reduction of area (which would never have been attempted but for the experience gained in intermittent filtration at Troedyrhiew) the charge upon the rate-payers of Merthyr, though still great in comparison with what it would have been, had the Troedyrhiew land only been utilized, will be reduced to less than half what it would have been had the original intention of 1869 remained in force, thus affording the most tangible evidence that could be produced of the true value of intermittent filtration, which in the first instance was regarded with derision, and which is still held up by its opponents as an object for public distrust."

It is to be remembered that at Merthyr Tydfil, as elsewhere in Great Britain, the sewage to be dealt with is largely increased in volume during storms, and that this increase imposes a great additional tax on the land devoted to its purification.

In filtration, even more than in irrigation, it is important to prepare sewage for application to the land, by the removal of its coarser parts and of its fibrous and adhesive elements, which would tend to clog the surface of the filtering area. There is no adequate advantage in resorting to chemical precipitation as a preliminary process, because all that can prove an obstruction to easy absorption can be much more cheaply and simply withheld in the manner indicated in the foregoing pages, especially in Chapter VIII.

The regulation of the land for intensified filtration is often a much more serious operation than anything that is called for in irrigation. In order to ensure maximum absorption and the equal use of the whole area, it is desirable to submerge, under a head of some inches, either the whole surface, or the trenches into which sewage is run between elevated beds. This requires careful and exact grading, and, if the whole surface is to be covered, the surrounding of the area with an elevated embankment.

To illustrate what is meant by careful and exact grading, reference may be made to the disposal field at the Insane Asylum at London, Canada, where, in grading a level section of about five acres, and in dividing it into beds and trenches for lateral absorption, we set about 4000 stakes to grade and line, 1000 to line only, and 1000 to grade only.

If the whole space to be devoted to filtration is not quite level, it may be graded to a terraced form, or otherwise divided into separate fields at different elevations. Each area must be level; but when there is a division into beds and trenches, each trench may be graded by itself along the contour of the slope, and may overflow at its end into the one next below it. In this, the current will be in the reverse direction, and it may overflow into the next, the flow being again reversed. The drop at each overflow will be more or less steep, according to the general slope of the land, and it must be protected against washing by suitable paving. In regulating this method of delivery, provision may easily be made for cutting out any trench or group of trenches, so that the alternate use of different sections may be controlled by gates at the end of each trench at one side of the field.

If the filtration is through the surface of the whole tract, and is so heavy as to prevent vegetation, or if the sewage is run in trenches, then it will be requisite, from time to time, at the end of a resting period and when the ground is dry, to rake it over and break up any hardening or caking of the surface. A fungus sometimes grows, which forms a sort of a mat or felting at the bed of a trench from which sewage is to be absorbed, but this cracks and curls up in drying and gives little trouble.

The uniform distribution of sewage over a flat area of considerable size will be sufficient from a single point of discharge, if the soil is slow enough in drinking it in, to cause the whole surface soon to become submerged; but more porous ground will need a discharge at different points along the borders, and perhaps even a delivery by outlets in the interior of the area. When absorption and transmission are very rapid, purification is less complete, and parts of the land not near the point of delivery may not be reached at all. Such a soil will be much improved by a thin covering of finer sand or earth, so applied as to check the too rapid sinking of the sewage into the ground. The more copious the flow, the more completely will it extend over the whole area, and a better distribution may be secured by accumulating the sewage in a tank, to be discharged from time to time through a large outlet.

Filtration is a more artificial process than irrigation, but it may, in some cases, be more desirable on the score of economy. In many instances the two will be used together; either an irrigation farm will have a filtration area for use in emergencies,— or more or less regularly,— and a regular filtration area will be worked in connection with land which may be flooded when it is desirable to give one of the filtration beds a longer rest than its regular turn would allow, or when the crops growing on this land would be benefited by irrigation.

As has already been indicated, the two methods of treatment are, in reality, only two degrees of the same process,— the intermittent application of sewage to land for the destruction of its organic impurities by bacterial oxidation, as set forth in the account of the Massachusetts investigation in Chapter IX.

CHAPTER XIII.

In the preceding chapters it has been shown that, in the disposal of sewage by irrigation or by filtration, the inoffensive destruction of putrescible organic matters is accomplished by the action of living organisms (bacteria), and that these organisms can work effectively only when a sufficient quantity of oxygen is present. In the absence of air, certain forms of putrefaction and disintegration will be set up, but such processes are offensive, and, to the best of our knowledge, they are usually dangerous.

It has also been shown that irrigation and filtration are identical in theory, — that purification is accomplished by the same agents (bacteria) in both cases. The difference between them is merely a question of degree, and the intensified efficiency of filtration is due only to a more careful provision and maintenance of conditions favorable to the action of the common purifying agents. In each of these methods, as well as in the case of the filters at Lawrence, complete and inoffensive purification is obtained only by the intermittent application of sewage. The liquid applied must be allowed to drain away and be followed in its descent through the interstices of the soil by the air necessary for the bacterial destruction of the impurities adhering to its particles.

In 1891 it occurred to the writer that conditions might be artificially produced more favorable to bacterial oxidation than those of intermittent downward filtration, and application was made for a patent covering a system of forced aeration of the medium used for sewage filtration.

The belief was that an abundant supply of oxygen in all parts of the filtering medium at all times must increase the activity of the nitrifying organisms by extending their field of operation, and that all the air needed for continuous and efficient purification might be introduced artificially from below.

It also seemed probable that the solid organic matters suspended in sewage could be strained out by passing the flow slowly through tanks filled with suitable filtering material ; and that, when this material had become choked, the accumulation of sludge could be destroyed and the filters restored to usefulness by draining them and forcing air through them, artificially stimulating bacterial oxidation.

To determine the value of these theories, an experimental plant was erected and put in operation at Newport, R.I., in 1894. The process consisted in the mechanical straining out of all solid matters carried in suspension in sewage, and their subsequent destruction by forced aeration, and the purification of the clarified sewage by bacterial oxidation of its dissolved organic matters in an artificially aerated filter.

The results accomplished exceeded the most sanguine expectations. Sewage loaded with grease, dirt, excreta, and the putrid overflow of cesspools, escaped from the tanks clear, white, and limpid. The impurities were not passed through in disguise, the foul smell was not masked by other odors; the effluent water was clean, — a good drinking water.

This complete regeneration continued through five months, and when the filters by which it had been performed were taken apart, they, too, were clean. The filth had not accumulated in them; it had completely disappeared.

Described briefly, the apparatus and the mode of operation are as follows : —

The sewage, after passing through suitable screens, which withhold large solids, such as rags, paper, lemon-rinds, etc., flows slowly, horizontally, through or over a shallow bed (say 6 inches deep) of coarse broken stone or similar material, which serves to catch and retain the coarser floating particles which have escaped the screens.

These broken-stone beds should be provided in triplicate, each to have ample capacity to receive the entire flow for a certain period ; and they are to be used in alternation, allowing to each twice as much time for rest and recuperation as for active service. When one of these areas is thrown out of use, it is drained and its accumulation of filth is exposed to the action of air, which results in its speedy destruction, leaving the bed in condition again to receive its quota of sewage when its turn comes.

Leaving this area of broken stone, the sewage, freed from its coarser solids, passes to a straining tank filled with fine broken stone, coarse gravel, locomotive cinders, coke, or similar porous material. This tank is divided into two compartments by a vertical diaphragm, which extends nearly to the bottom of the tank. The sewage passes down through one of these compartments, flows under the diaphragm, and rises through the other compartment, overflowing at its top. The rate of flow through the tank must be sufficiently slow to allow the deposition upon the surfaces of the filtering medium of the solid particles suspended in the sewage. If the speed be properly regulated, practically all of the suspended impurities are retained in this tank, and the sewage leaves it as a slightly opalescent but clear liquid, with a perceptible odor. At this stage it compares favorably with the effluent of chemical precipitation works.

When one of these straining tanks has been in operation for a considerable time, the accumulation of sludge at the surface of the filtering material clogs the pores of the filter and decreases its capacity, although the quality of the effluent is in no wise impaired. When this condition is reached, the flow is turned to another tank of similar construction, where the straining process begins anew. The filter tank which has just been thrown out of use is drained, and an abundant supply of air, under light pressure, is forced by a blower into the bottom of the tank (where means for its even distribution are provided), rising through the filtering medium in a strong current, which penetrates into all its voids and pores. Under these conditions, rapid bacterial oxidation is set up, and the retained impurities are speedily consumed, leaving the tank in a clean condition, ready for further use.

It will probably be best to provide four of these straining tanks, to be used in alternation, allowing to each a period of aeration three times as long as its period of use. The filters which were in use at Newport during the whole time of the experiment (over five months) without any renewal of material, showed no signs of deterioration, but were practically as clean at its termination as when they were new, and were capable of producing as good results.

The degree of purification attained at this stage of the process is, in many cases, sufficient to satisfy all requirements. When purification to a drinking-water standard is necessary, it may be obtained by further treatment as follows : —

After passing the straining tanks, the sewage, which has been relieved of all matters in suspension, but which still contains nearly or quite all of the dissolved impurities originally in it, flows to an aerating tank. This is similar in construction to the straining tanks, save that it has no dividing diaphragm, the sewage passing in at the top and escaping, as purified water, through a trapped outlet at the bottom. It must also be considerably larger than the straining tanks, for the sewage, instead of passing through the filter in a solid column, as in the former case, trickles down in a thin

L

film over the surfaces of the particles of coke or other filtering material ; while through the voids between the particles, and in immediate contact with the trickling films of liquid, a current of air is constantly rising, being introduced at the bottom of the tank by a blower.

After the filter has been in use a short time, the nitrifying organisms which have entered with the sewage, develop, and, in the presence of abundant food supplied by the sewage and abundant oxygen furnished by the blast, multiply with great rapidity, until their number has reached a point at which the average food supply is only capable of feeding the existing colony, and further multiplication is checked. When a sufficient colony of organisms has become established, the consumption of the organic matter in the liquid passing through the tank will be practically complete, so long as the quantity is reasonably uniform. Any sudden and marked increase or diminution in the rate of flow will make it necessary for the colony of organisms to adapt itself to the new conditions, and this it will do, within reasonable limits, in a short time.

As these nitrifying tanks are constantly aerated, *they are used continuously*, and one area, of sufficient size to care for the flow, is all that need be provided.

The process by which the impurities of the sewage are removed is the purely natural one on which depends the ultimate destruction of all organic matter. When sewage is spread over the surface of the ground, as in irrigation, it is exposed to the atmosphere in thin broad sheets, and the bacteria which reduce its putrescible matters are active because air is abundant. The process in the aerating tank described above is essentially the same, but in this case the earth is massed in cubical form, and the atmosphere is made to pervade the mass, so that every conceivable plane within it presents — so far as bacterial activity is concerned — the conditions of a natural surface.

The same is true with regard to the straining tanks. While the sewage is passing through them, the action is merely mechanical sedimentation. When the liquid has been drained off and the aeration has begun, the process and the result are the same as they would be if the accumulated sludge were spread in extremely thin sheets over the surface of a large area of soil.

The operation of the experimental plant and the results accomplished are set forth in detail in a report published after the close of the test.[1] It is sufficient for the purposes of this chapter to state that the average daily flow through the strainers was at the rate of 7,574,400 gallons per acre (the maximum was 17,900,388 gallons), and through the aerators at the rate of 1,064,213 gallons (the maximum, after nitrification began, being 4,826,112 gallons).

The average percentage of purification, as represented by the removal of organic nitrogenous matter,

---

[1] This pamphlet, which includes the report of Mr. George W. Rolfe, the chemist in charge, and complete analytical tables, will be sent, without cost, to readers who desire it, upon application to the author at Newport, R. I.

accomplished by the strainers alone, was 51.2, and by the strainers and aerators together 92.5. At one time a purification of 99.08 percent was reached.

When the tanks were taken apart at the close of the experiment, the upper foot (approximately) of the receiving compartment of each of the straining tanks showed more or less accumulation of silt, probably the result of the few heavy rainfalls, during which pumping was continued, bringing much gutter mud to the tanks. Below this, the material was apparently as clean as when first put in, the pebbles and white gravel looking as though they had just been taken from their native beach. In no part of the tanks was there any sign of organic matter or any suggestion of the hundreds of thousands of gallons of sewage which had been passed through them. The thin layers of sand on top of the aerators were black with sulphides, but all the material below this was sweet and clean. No impurities had been *stored* in any of the tanks. They had been detained and destroyed. All the conditions clearly indicated that the usefulness of the filters had become in no wise impaired, that they were capable of performing their functions indefinitely, and that, under proper management, no renewal of the filtering medium would be necessary.

Briefly summarized, the results of the experiment demonstrated that : —

1. The suspended matters of sewage (sludge) can be mechanically withheld by straining slowly through suitable material.

2. The filth accumulated by this straining material can be destroyed and the straining medium restored to a clean condition by mere aeration.

3. The successive alternate operations of fouling and cleansing can be carried on indefinitely, without renewal of the straining material.

4. The purification obtained by this straining process practically equals that accomplished by chemical precipitation, and is sufficient to admit of discharge into any considerable body of water not used as a source of domestic supply or for manufacturing purposes requiring great purity.

5. Practically, all of the dissolved organic matter in sewage can be removed and purification to a drinking-water standard can be obtained by the use of suitably constructed bacterial filters.

6. Such filters can be maintained in *constant and efficient* operation by suitable aeration.

7. The erection of a plant capable of purifying large volumes of sewage upon a relatively small area calls for no costly construction. Repairs and renewals are merely nominal. The attendance required is but slight. There is no outlay for chemicals, etc. The only expense of mechanical operation is the driving of the blower or air-compressor.

8. The process admits of wide variation in the selection of filtering material, and nearly every community can find, in its local resources, something suitable for the purpose.

CHAPTER XIV.

CHEMICAL TREATMENT.

UNDER this head there may be grouped all processes which seek, by the addition of chemical reagents or absorbents, to remove the impurities of sewage, in whole or in part, or to render them less obviously offensive. The object and result of this treatment, determined by the conditions to be met, may be in the direction of either deodorization, disinfection, mere clarification, or actual purification. Any or all of these results may be obtained, to a greater or less degree, by the use of suitable chemical methods.

Concerning mere deodorization, it may be said that, with a properly constructed and well-managed system of collection and disposal, sewage should never need deodorizing. It should cease to exist as sewage before its latent noxious properties can be developed. Fresh sewage is practically odorless, and it should not be allowed ever to become otherwise. When deodorizng processes are applied to the foul outflow of badly planned or badly constructed sewers, any resulting improvement of condition is apt to be apparent and temporary. Efforts were made by the Metropolitan Board of Works (London) to deodorize the water of the Thames, made foul by the discharge of sewage at Crossness and Barking. Chloride of lime (bleaching-powder) was first tried, but the results were unsatisfactory, and its use was abandoned. Some of the alkaline manganates and permanganates were substituted, but even the apparent results were by no means commensurate with the cost of their application. Without the subsequent use of a neutralizing acid, the alkaline quality imparted to the sewage by these chemicals would, under some conditions, stimulate putrefactive changes.

Disinfection of sewage, being the destruction of the organisms naturally existing in it, is proper and desirable only when it results from the destruction of the organic matter on which such organisms feed. When this destruction is complete, the organisms will disappear. Until then the sewage should remain infected, and subject to such influences as will facilitate the work of the bacterial agents of inoffensive decomposition. In other words, the organisms should be starved out, not poisoned. Even with the most thorough process of forced disinfection, reinfection and subsequent decomposition are inevitable — inoffensive and innocuous, or offensive and dangerous, according to the conditions. In the former case nothing has been gained by delay; in the latter case much has been lost.[1]

Neither complete purification, nor a very near approach to it, can be effected by any practical chemical treatment known; but a high degree of clarification may be reached

[1] This does not apply to the specific disease germs which exist in excreta during certain infectious diseases. Specific immediate disinfection is here called for; but this is in the province of the nurse, not of the engineer.

and a not inconsiderable amount of purification may be accomplished by processes of chemical precipitation. The elaboration of these processes dates from 1762, when a patent for the chemical treatment of sewage was taken out by Deboissieu. Since then patents have been granted by the hundred, for methods good, bad, and indifferent. Most of these may be disregarded. The more important ones will be considered in due course.

Precipitation was the outcome of unsatisfactory sedimentation. If the flow of a liquid carrying suspended matters be checked, as by detention in subsidence tanks, the heavier particles will gradually sink to the bottom and form a deposit, and the lighter particles will float to the surface. With sewage, which contains many substances of widely different gravity, this process is slow and inefficient; but if suitable chemicals be added, these combine with certain constituents of the sewage, forming an insoluble coagulum, which falls to the bottom, carrying with it the suspended matters and, under favorable conditions, a portion of the dissolved impurities. The deposit is known as "sludge." This is, practically, the only chemical part of the process. The mixing and distribution of chemicals, the decanting of the clarified water, and the collection and disposal of the sludge are mechanical operations, which form a troublesome and costly part of the treatment.

The results to be sought, stated in the order of their importance, are as follows : —

1. To produce an effluent coming fully up to the predetermined standard of purity.

2. To accomplish this with a minimum outlay for construction, chemicals, and operation.

3. To keep the quantity of sludge at a minimum.

4. To produce a sludge that will part with its water readily, that will keep without odor or offence, and that will contain a maximum of the manurial elements of the sewage.

Certain conditions and details of treatment are essential to the success of chemical precipitation. They may be stated thus: —

1. The sewage should be fresh; that is, it should have undergone no putrefactive changes. Such sewage will yield a better result, with a smaller amount of chemicals, than a sewage in which any breaking up of the constituents has begun.[1] For this reason, as well as for the purpose of ensuring a uniform quantity[2] and

---

[1] The decomposition of the urea should be excepted. This begins almost immediately, and speedily results in the formation of carbonate of ammonia, which is harmless. This change is without offence, and a considerable time elapses before any other decomposition occurs.

[2] Baldwin Latham, in discussing Dr. Tidy's paper on the "Treatment of Sewage" before the Society of Arts, May 5, 1886, said that if the rain were not excluded from the sewers, "the average daily flow might be increased twenty to thirty times in one hour by a heavy shower," and that this was "important in a chemical plan, simply because to carry out such a process properly you required large tank space, and if in a limited time thirty times as much rain as the ordinary volume of sewage came down, as the tanks generally held only six hours' sewage, instead of having six hours for precipitation, there would only be a few minutes, and the consequence was the sewage must go away in a polluted form. Sufficient tank room could never be supplied (unless the works were on an enormous scale) without the surface water was excluded." — Journal of the Society of Arts, p. 666.

quality of sewage to be treated, the flow from a strictly separate system of sewers is to be preferred.

2. The sewage should be passed through suitable screens, or other straining device, to arrest the coarse solids and miscellaneous floating objects contained in it, which are not susceptible of precipitation, and which must be withheld from the effluent.

3. The chemicals should be added in sufficient amount to effect complete precipitation, and they should be thoroughly mixed with the sewage before it flows into the settling-tanks.

4. It is important that abundant tank room be furnished, so that ample time may be allowed for the complete subsidence of the precipitate, and so that each tank may, in its turn, be frequently thrown out of use and cleaned.

5. In cleaning, the sludge should be completely removed, and the tank itself thoroughly washed down before being put in use again.

6. Good management, especially as regards a scrupulous cleanliness of the whole plant, sound judgment, and constant supervision are absolutely essential to success.

7. It is advisable that frequent chemical and biological analyses should be made of both sewage and effluent. In this way only can uniformly good results be maintained with certainty.

The general operation common to practically all precipitation processes may be described as follows: —

Preferably, the grosser solids are removed by screening or straining, and this may be done in any of the methods indicated in Chapter VIII., or by passing through filter-tanks in which walls of coke or similar material act as strainers. Dr. Tidy, however, in his paper on the "Treatment of Sewage,"[1] indicates that it is not desirable to remove anything beyond the coarsest substances, as "the precipitation appears to be more complete as the quantity of suspended matter in the raw sewage increases." This, of course, involves an increase of sludge.

After passing the screens or strainers, the sewage, on its way to the settling-tanks, receives its quota of chemicals. These have been prepared by dissolving them, if they are soluble, or by grinding and intimate mixture with water if they are insoluble, and they are fed into the sewage by suitable devices, care being taken to ensure a thorough mixture before the settling-tanks are reached.

The list of reagents proposed is a long one, and includes many chemicals and compounds of chemicals which may be rejected without serious consideration, either because of their inability to purify the sewage to the required degree, because they produce an effluent chemically injurious, or because of prohibitive cost. Following this principle of "choice by rejection," investigation as to the comparative merits of various processes is much simplified. In selecting the process to be used, several important considerations present themselves.

[1] Journal of the Society of Arts, p. 1171.

The requirements of the case must be fully satisfied by the production of an effluent which shall at no time fall below the standard of purity adopted, and no process which results in a less degree of purity should be considered, even although, in other directions, it may seem most desirable. This adopted standard of purity, if properly determined, indicates the lowest permissible point of purification, both organic and chemical. It is the result of a careful study of the conditions which may be affected by the output of the works, including not only hygienic considerations, but also considerations of business and of sentiment.

If the effluent is to be delivered into a stream which, before receiving it, is fit for domestic use, it must be so purified before discharge that this fitness shall not be impaired. If the stream be not used for domestic purposes, but serve for watering cattle, though the demands are less exacting, a high degree of purification is still required. If the water is to be used for manufacturing purposes, perhaps a satisfactory clarification may be all that is needed; but for use in some of the arts where great purity is necessary, even the chemical composition of the water must at times be taken into account. If, for example, soft water be needed, and the natural flow of the stream furnishes this, we must not make the water hard by discharging into it the effluent from a lime process. Even the uses of water for pleasure must be considered. Clear water must not be made turbid, and odorless water must not be made offensive. In short, there should be no reason for any unfavorable

criticism of the process on the part of any one who is to
be affected either by the effluent itself or by conditions
caused by it. The legal rights of riparian owners are
set forth in Chapter XV.

The probable effect upon fish is also to be considered.
Fresh sewage, from a system that receives no poisonous
manufacturing refuse, is not only harmless to fish when
sufficiently diluted, but it contains matters which they
eat greedily. The introduction of certain chemicals,
however, which are used as precipitants of sewage, may
exert a very destructive influence upon them, and where
the continued existence of fish in a stream about to
receive the outflow of a precipitation plant is desired,
special consideration must be given to the composition
of the effluent. Among the reagents most destructive
to fish life rank solutions of free chlorine and of the
hypochlorites (chloride of lime and bleaching-soda).[1]
Caustic lime, whether used as quick-lime, slacked lime,
or lime-water, is destructive to fish. Slater says concerning it:[2] "I should advise riparian owners, lessees of
fisheries, etc., to protest against its introduction into

[1] The experiments of Saare and Schwab proved that solutions containing 0.04 to 0.005 of one per cent were fatal to tench, and that a
solution of 0.0008 was deadly to trout. Placed in this, they soon
turned on one side, and even removal to fresh water did not restore
them. The presence of an acid increases the destructive effect of
chlorine compounds. The Royal Rivers Pollution Commission pro-
tested "against these substances as unfitting the water of rivers for
almost every conceivable purpose, and especially rendering it deadly
to fish."

[2] J. W. Slater, Sewage Treatment, Purification, and Utilization,
p. 20.

their waters. One manner in which lime present in waters destroys fish is by entering their gills, and being there precipitated by the carbonic acid exhaled, it forms deposits of carbonate of lime, which interfere with respiration."

The salts of iron and the salts of aluminum, in any considerable quantity, are also poisonous to fish, but, as the proportion of these reagents used in precipitation processes can be so closely regulated that little if any escapes in the effluent, practically no danger is to be apprehended from their use under careful management.[1]

Of course the degree of injury done to fish by chemicals of a poisonous nature will depend upon the quantity escaping in the effluent, and upon the volume of the diluting flow in the stream receiving it. With abundant dilution, no bad results may be apparent; but Saare and Schwab, in summarizing the results of their experiments, "declare every substance soluble in water to be more or less injurious to fishes. Proportions which do not produce acute disease will probably be found hurtful on more prolonged action, and will especially interfere with the multiplication of fish."

---

[1] Certain salts of lead are capable of precipitating both suspended and dissolved organic matter, but their presence in a stream would be fatal not only to fish and cattle, but also to vegetation. Salts of barium will precipitate carbonic and phosphoric acids and will remove any free or combined sulphuric acid contained in water, but any excess passing out with the effluent would prove poisonous to both animal and vegetable life. And further, any remaining in the sludge in a soluble condition would be detrimental to vegetation if applied to land. Zinc cannot properly be used in treating sewage, because of its poisonous qualities, though many processes prescribe it.

Second to the question of efficiency and safety, in the selection of a suitable reagent, comes the question of cost. The precipitant "must be cheap — cheap not merely by reason of present small demand, but of abundant, or rather unlimited, supply." Numerous compounds, which give excellent results in the laboratory, are unavailable for use on a large scale, because of their excessive cost.

Another consideration which should affect the choice of a chemical is that it is desirable to keep the amount of sludge at a minimum, and if there is any chance of recovering a portion of the cost of treatment from its use or sale, to make it of the highest possible manurial value. This being done, we either have a sludge of the maximum value to sell, or, if unsalable, we have less troublesome matter to dispose of. For this reason, solutions are generally preferable to solids. When lime is the only precipitant employed, the sludge is bulky and of little manurial value, making its disposal difficult, and precluding its profitable use as a fertilizer. When salts of aluminum are used, the amount of precipitate is but slightly larger than the amount of solid material withdrawn from the sewage, and the resultant sludge is of considerably greater value.

Nearly all the ordinary processes of precipitation remove the phosphoric acid, an important constituent of urine, and invariably found in sewage. The removal of this is of great advantage; for, although not present in organic form, it is objectionable as favoring the growth of microscopic organisms. Its addition to the sludge increases the manurial value considerably. Free ammonia and the alkaline nitrates and nitrites cannot be precipitated by any practical method now known. Their escape is unimportant when the purity of the effluent is considered, but their retention in the sludge would add largely to its value.

As a matter of secondary importance, it may be said that certain precipitants impart a distinct color to the effluent, or to substances with which it comes in contact. Prominent among these are the iron compounds, which are otherwise valuable. "Waters to which they have been added take a greenish-yellow color on prolonged exposure to the air, and a yellow, ochreous deposit is formed on stones, brick-work, piles, etc. Though these deposits may be perfectly harmless, yet to the public they convey the notion of an excrementitious origin, and the process is at once condemned. Iron sediments, if containing sulphur, or if coming in contact with sulphur compounds (mineral or organic), turn intensely black, and have an unpleasant appearance."[1] This objection does not seem to be a serious one. It is purely sentimental, and the abundance of natural iron waters producing similar discoloration should afford ample evidence that its appearance is in no way due to sewage contamination. The matter is noticed by Mr. Hazen, of the Massachusetts State Board, in his report on the experiments with chemical precipitation at Lawrence, but he does not consider the color necessarily objectionable.

[1] J. W. Slater, Sewage Treatment, Purification, and Utilization, p. 89.

An ideal precipitant would be one which removed all matters suspended in the sewage, and all putrescible matters in solution, together with all disease germs and putrefactive bacteria, yet which would not destroy the oxidizing and nitrifying organisms which tend to complete the operation of purification; which retained in the sludge all substances needed by the soil — nitrogen, potash, and phosphoric acid; which exhausted itself in the operation, so that no trace of it remained in the effluent; which cost nothing, or next to nothing, and which could be applied as an aqueous solution, or, better still, as a gas, so that the sludge would be of minimum bulk. This ideal will probably never be reached, but it must be remembered that the subject is still in a state of development, and improvements in methods and results are constantly being made.

In the mean time, the selection of the chemicals must be made in accordance with the peculiar conditions of each case. They should be effective, safe, and cheap, and, preferably, capable of application in solution.

The value of any chemical precipitant depends largely upon the care and skill with which it is prepared and the thoroughness of its admixture with the sewage. With inefficient mixing, the amount of precipitate, and, consequently, the degree of purification, is less than could be obtained with the same material under more careful management. In addition to this, a portion of the reagent will escape combination and be wasted, flowing off with the effluent in its free form, possibly with results injurious to aquatic life, or to manufacturing processes.

If the chemical is to be used in the form of a solution, care should be taken to secure the dissolving of all that is introduced into the mixing-tanks, and to maintain a uniform strength of solution. When the sewage flows to the subsidence tanks by gravity, the chemicals may be introduced into the channel leading from the outfall, either by permitting them to flow from a trough placed across the channel, and notched on one side, so as to allow the solution to escape in a number of small streams, or by delivering it through a pipe laid across the channel and perforated with a number of holes. Either of the above methods will ensure a wide distribution of the liquid, but some further device had better be employed to produce more intimate mixture of the sewage and the precipitating agent. This may be accomplished by the use of mechanical stirrers, or of a mixing-wheel driven either by power or by the flow of the sewage itself. Or a good result may be obtained by converting a portion of the channel to the subsidence tanks into a salmon-ladder, by placing baffle boards in it projecting alternately from one side and the other. In cases where it is necessary to pump the sewage to the precipitation tanks, the chemicals may be introduced into the pump-well, the action of the pumps producing a thorough admixture.

As the volume of sewage varies from hour to hour, it is necessary to adjust the amount of chemicals to the ever-changing conditions. Unless some automatic device for this purpose be adopted, constant watchfulness on the part of attendants will be necessary. But automatic control can easily be provided by means of floats, which, rising or falling with the varying level of the sewage, open or close the valves through which the chemicals are admitted. Or mechanical regulators can be used, driven by a water-wheel placed in the sewage channel. The quantity of chemicals supplied, being governed by the speed of the water-wheel, will be proportional to the flow of sewage. A change in quantity of flow, however, does not always indicate a change in the amount of chemicals required, for increase of flow may mean merely increase of dilution. On the other hand, the volume of flow may remain practically constant, while the strength of the sewage increases or decreases materially. No mechanical device can properly regulate the addition of chemicals under such conditions, and careful attention and sound judgment must be relied upon where the maintenance of a uniform effluent is important. As these changes in the composition of sewage are rapid, it is obvious that no process of chemical analysis can be used in determining the treatment for the time being. Slater suggests the following method,[1] which is simple, and, as a rule, sufficiently accurate: —

"Suppose that we are using two important ingredients, *a* and *b*. We take four small hydrometer glasses and fill them to a known height with the sewage *after* it has received the mixture. Number 1 glass we leave

[1] Sewage Treatment, Purification, and Utilization, p. 114.

as it is.   To No. 2 we add a little more of *a* (a solution
of which is kept ready at hand for the purpose); to No.
3, a little more of *b*, which is also kept ready; and to
No. 4 a little more raw sewage.   We then observe which
of the four glasses is the best, *i.e.* which goes down
with the cleanest, boldest flakes, leaving clear, colorless, inodorous water above and between.   If No. 1 is
the best, the treatment is continued as it was; if No. 2
is seen to be an improvement, we increase the proportion of *a* added; if No. 3 has the advantage, we
increase in like manner the dose of *b*.   And if No. 4,
to which more raw sewage has been added, is the best,
we see that we have been using too much material, and
we reduce the quantity accordingly. . . .   There are
other methods by which the treater may be guided.   He
may, from time to time, dip up a hydrometer glass full
of the treated sewage from different points of the channel . . . and set them aside in a good light to observe
how they settle.   He should occasionally walk round
his tanks and inspect them from different points of
view as regards the light.   If any process is acting well,
the effluent water, when in bulk, appears of a peculiar
bluish tinge.   This indicates not merely 'clarification,'
but 'purification,' since this blue tint, according to the
researches of Mr. W. Crookes, F.R.S., is the more decided the freer a water is from dissolved organic matter.
The treater must learn, by careful, intelligent observation, what modifications the sewage he has to deal with
generally undergoes; at what hours, on what days, or
under what conditions of weather, etc., these changes
come on, and how they are all to be met."

Warm sewage is more readily precipitated than cold. If the precipitants be heated before their addition to the sewage, they will combine more easily with its impurities, but the cost would probably be too great, and it would add another element to a complicated process.

The chemicals having been added to the flow in proper proportion, and thoroughly incorporated with it, the treated sewage passes to the subsidence tanks, where its velocity is checked or completely arrested, and the precipitated impurities gradually settle to the bottom, leaving the water clarified and more or less purified. This process of deposition may be either continuous or intermittent. In the former, the sewage continues its constant motion, though at a greatly reduced velocity, and the flow only ceases when a tank is to be cut out of operation and cleaned. In the latter, each tank is filled in turn, and its contents allowed to stand undisturbed until all the precipitated matters have settled to the bottom, the flow in the mean time being turned into other tanks.

Two methods of continuous precipitation are in use. One passes the treated sewage through long, flat basins at a very low velocity, the suspended matters gradually settling to the bottom, an equal amount of clarified effluent constantly escaping. Much depends upon the size and form of these tanks. Their capacity should be carefully calculated, provision being made for the retention in the tank of the maximum flow of sewage for at least two hours — preferably for as much as four, or six, or even eight hours. An extra tank must also be provided, adding, say, one-third to the total capacity, in order that each tank may in turn be thrown out of use and cleaned; the whole series being so arranged that any tank may be thus cut out for repairs or cleaning without interfering with the regular and efficient operation of the others. The tanks should be neither so shallow as to subject the precipitated matters to the effect of the surface current, nor so deep as to require an inordinate time for them to reach the bottom. A depth of five or six feet will satisfy both of these conditions. The treated sewage should be given the longest practicable journey in flowing from inlet to outlet, but the channel must not be so restricted as to cause an increase of velocity, which would tend to retard the deposition, and especial care must be taken to prevent the formation of eddies or rotary currents, which tend to "bore up" the mud from the bottom. In some cases the tanks are divided by partitions, rising to a point just below the surface, over which the sewage flows in thin sheets, the suspended matters below the surface being arrested by the wall, and slowly settling without further progress. The utility of these sunken partitions, however, is doubtful, for the velocity of the sewage flowing over them is increased, and this may induce an upward current of the deeper strata and produce the result which we seek to avoid. If the depth of flow is uninterrupted, the section of moving water is at the maximum, and the velocity consequently at the minimum. This condition is probably the best adapted for speed and thoroughness of deposition. The walls should be built of masonry or concrete, and smoothly plastered. Special protection against frost is not necessary, as the sewage is warm enough to prevent trouble from this cause. The bottom should be of brick or concrete, laid with a fair fall from end to end, and from the sides to a sludge channel in the centre, leading to a sump, from which the sludge can be pumped or drained away after the clear liquid has been drawn off. The inlet should be so arranged that the sewage enters the tanks quietly, without agitating their contents, and the outlet should be ample to prevent an increase of velocity as it is approached. In some cases, where improvement in appearance of the effluent is all that is sought, and where chemicals are used in indifferent quantities, with indifferent care, — and with indifferent results, — floating scum-boards are placed in the tanks to catch and retain froth, scum, and floating objects; but where the sewage has been properly strained and thoroughly precipitated, these are not needed.

It is often urged that sewage tanks should be covered: first, to prevent changes of weather from interfering with the process; and, second, to prevent the escape of odor. Such roofing is not only unnecessary; it is not desirable. The temperature of sewage as it flows from the outlet is rarely below 40° Fahrenheit, and when massed in large tanks it retains its heat for a considerable time. As for the odor, there should be none of any moment. If the tanks are ill-smelling, it is because the process is inefficient or the management incompetent or neglectful. The free admission of light and air is of no little benefit in furthering purification.

Another form of tank for continuous chemical precipitation is the "upright" or "vertical," which originated in Germany, and which is there used in a variety of forms, differing slightly in details, but essentially the same in principle. These tanks are usually circular, of considerable depth, and furnished with funnel-shaped bottoms. Treated sewage is introduced at or near the bottom of the tank and rises slowly to the top. The upward movement of the liquid is, however, slower than the rate at which the precipitate descends, so that the latter accumulates at the bottom, where it can be drawn off or pumped out, while the clarified water overflows at the top. The action of the precipitate in these tanks is curious, and is thus described by Mr. Rafter:[1] "It is found in practice that there is, in vertical tanks, what may be termed a neutral plane of precipitation. Any organic matter which may happen to pass above this plane, as it is more thoroughly acted upon by the chemicals, slowly falls back, in opposition to the upward current, to the neutral plane. The flocculent matter collecting there forms a sort of filtering medium, which assists in arresting other matter which is floating upward. When a considerable mass has collected, the whole finally falls to the bottom, and the process of collection at the neutral plane again takes place." The various modifications of the process include ingenious means for distributing the sewage and chemicals

---

[1] Sewage Disposal in the United States, Rafter and Baker, p. 207.

throughout the tank, for collecting the effluent in overflow channels (so that there is no rush of water to a
single point of outlet), and for removing the sludge.
The main advantages claimed for the method are —

1. The small area required for tanks of large capacity.

2. The ability to remove the sludge at any time without interfering with the regular working of the tank.

3. The saving in tank capacity, due to the fact that
a tank need not be thrown out of use for cleaning.

This form of tank was used at the sewage disposal
works of the World's Fair at Chicago, where the result
produced was by no means such as to commend it.

Intermittent precipitation requires tanks of practically
the same form and size as the process of continuous
precipitation. Their arrangement should be such as to
furnish a free and independent discharge for each tank.
The time needed for deposition under this method is
somewhat less than in the case of continuous precipitation, but the actual capacity is practically the same, as
the time spent in filling and emptying must be added
to the time needed for deposition. The speed with which
the impurities of well-precipitated sewage subside when
at rest is thus stated by Dr. Tidy:[1] "The water will
begin to clear a few minutes after the cessation of agitation. In thirty minutes it clears to a depth of three
feet with eight inches' precipitate, while after two hours
the precipitate will measure four and a half inches only."

Decision as to the respective merits of the three forms
of precipitation tanks described above is difficult. At

[1] Journal of the Society of Arts, p. 1182.

first sight, it would seem that the intermittent process, which brings the treated sewage to absolute rest, would be the most favorable for the complete subsidence of precipitated matters, and therefore to be preferred, but there are objections to its use, under some circumstances, which must not be overlooked. As soon as a tankful of sewage has settled, the supernatant water is drawn off, leaving the sludge in the tank. Where there is a gravity discharge to a common sludge-well outside of the tank, this may be run into it; where there is no such provision, it must either be pumped out or allowed to remain. Even if the sludge be drawn off after each emptying of the tank, the sides and bottom will remain foul, unless cleaned before the sewage is readmitted, and when the flow is again turned into the tank, the foulness is stirred up and mingled with the fresh sewage, making its clarification more difficult and more liable to be offensive. To clean a tank each time that a tankful of sewage is discharged would be expensive. Furthermore, tanks for intermittent use require a considerable grade from inlet to outlet, in order that they may be emptied speedily and thoroughly. In very flat districts this might prove a disadvantage.

In a system of continuous precipitation, the sewage flows quietly and slowly through the tanks, and if the distance traversed be long enough, and the velocity of flow kept low enough, practically all of the precipitated matters will subside. The sludge meanwhile gradually accumulates at the bottom, settling continually into a more and more compact mass, and if the tank be not too shallow, entirely free from any disturbing influence, until the time comes when cleaning is desirable. Tanks for continuous precipitation do not require any considerable grade within themselves, and the process may therefore, where but little fall is available, be preferable to intermittent precipitation.

Vertical tanks, as has been said, require less space than the ordinary shallow tanks, and this is an important consideration when a precipitation plant is to be erected in or near a city, where land is expensive. They have also the advantage that the sludge can be removed frequently (and consequently in the best condition for further treatment and ultimate disposal) without interfering with the operation of subsidence. It has been found, however, that sludge accumulates on the slope of the funnel-shaped bottom, putrefies there, and fouls the effluent. To obtain the same completeness of precipitation the flow through a vertical tank should be somewhat slower than through a shallow horizontal tank; for in the former, the rising current directly opposes the downward motion of the precipitated matters, while in the latter the influence of the horizontal current is practically *nil*, the sediment settling with almost the same speed as in a liquid at rest. The indications are, that to secure the best results, the average rate of flow through an upright tank should not exceed twelve or fifteen feet per hour.

Whether precipitation be intermittent or continuous, it is necessary to the securing of successful results that the tanks be frequently and thoroughly cleaned. If this be neglected, the treatment, however skilful in other respects, will avail but little. Each tank should be cleaned not less frequently than once in three days, or, better still, every other day. In continuous precipitation, when the flow is run through a series of tanks, the one which first receives the sewage after the admixture of the chemicals will accumulate the largest deposits, and will therefore require frequent cleaning. While the last tank of the series will contain the least deposit, its determining influence upon the purity of the effluent is, however, such that, though the accumulation is less, the cleaning must be frequent and complete. If the sludge be allowed to remain at the bottom of the tanks, it soon begins to putrefy and to give off offensive gases, which rise to the surface in bubbles, interfering with the settling of freshly precipitated matters, and partially neutralizing the effect of the chemicals which have not yet completed their work. As the putrefaction progresses, the sludge becomes saturated with gas, which eventually buoys up masses of it to the surface. As soon as the gas escapes, these putrefying matters sink again, fouling the contents of the tank and inducing further putrefaction. The most offensive constituents of sewage are those which are held in solution. Such of these as combine with the reagent in insoluble form constitute the lightest of the precipitated matters, and therefore settle near the point of discharge. These matters, being the most putrescible, are the most likely to rise again in the manner above described, and being already near the outlet can readily refoul an effluent which has once been sufficiently purified. If a good result is required, the necessity for clean tanks is imperative.

In continuous precipitation, the effluent water escapes from the last tank, over a weir, in a constant stream, of practically the same volume as the flow of sewage entering the first tank. In intermittent precipitation, means must be provided for drawing off the water (after all sediment has subsided) without disturbing the sludge accumulated at the bottom of the tank. This may be done by successively opening outlet gates placed at different depths, like the gauge-cocks of a boiler, or through a drain-pipe fitted with a flexible hose terminating in a funnel opening upwards. This can be gradually lowered by ropes and pulleys as the level of the water descends, until the mouth of the intake is just above the level of the soft sludge. A satisfactory method of running off the clear water is by means of a movable pipe, supported by a float, which keeps the mouth of the intake submerged a few inches below the surface. The rapidity of removal is regulated by a valve on the main drain.

The sludge which is left is allowed to run, or is pumped, into suitable receptacles, for immediate use or destruction, or for further treatment. It is of a thin, pasty consistency, practically odorless, if it has been precipitated by a good process, and if frequently removed in a fresh condition. It contains about 90 per cent of water, which makes its handling by gravity or by pumps easy and inexpensive. The amount of sludge produced will vary not only with the amount and strength of the sewage, but also with the method and means of precipitation, and the completeness or incompleteness of the purification. It is also influenced to a certain extent by the presence of some constituent not usually found in sewage, as when certain manufacturing wastes are discharged into the sewers, and by the admission or complete exclusion of surface water. The theoretical quantity of sludge can be approximately calculated from the amount of suspended matter contained in the sewage, and the chemicals used in precipitation, but the result is not always accurate. English experience shows that the average amount of liquid sludge (90 per cent water) precipitated from fresh sewage of water-closeted towns is about two and one-half cubic inches per gallon. The proportion in this country, where more water is used, is probably about one and one-half cubic inches.

This liquid sludge, as drawn from the precipitation tanks, may, without further treatment, be pumped or allowed to run on to agricultural land, for use as a fertilizer;[1] or, where a sufficiently large and deep body of water is at hand, it may be loaded into scows, carried to a suitable point, and dumped.

---

[1] A considerable amount of sludge may be disposed of on a comparatively small area by running it into shallow parallel trenches close together, the sludge, as fast as each trench is filled, being lightly covered with the earth dug from the next trench. After standing a year this field can be profitably cultivated, the sludge aiding the crops materially, and usually after two years' cultivation, *i.e.* in the third year after the previous application, the disposal of the sludge in trenches may be repeated.

Unless it can be disposed of by one or other of these methods, artificial means must be resorted to for more or less completely separating the solids and liquids of which it is composed, in order that its bulk may be reduced to a minimum, and that its capacity for putrefaction — which is greater in wet organic matter than in dry — may be lessened. The surplus water may be removed by any one of a number of methods. It may be run into tanks with floors of porous earth, suitably drained, where it may lose, by evaporation and drainage, about 50 per cent of its moisture in twenty days. When this condition is reached, the sludge is stiff enough to be easily handled, carted away for application to land, or stored for future use. A longer exposure, under favorable circumstances, may free the sludge of as much as 80 per cent of its moisture.

The excess of water may also be removed by a process of artificial evaporation, and the product either utilized as manure or burned in a suitable furnace. Where the sludge is the result of a process of precipitation by lime, it may be dried, calcined, and mixed with clay, forming a hydraulic cement of indifferent quality.

The partially dried sludge is sometimes mixed with loam, marl, leaves, stable refuse, etc., forming a compost heap which is ultimately used as manure. Where all the refuse of a community is systematically cared for by the authorities, this may be the best method of disposing of the sludge.

Generally speaking, however, the treatment of sewage sludge by the filter-press is more satisfactory than any other method yet devised. By its use, 50 per cent of the contained water may be immediately expelled, greatly reducing the bulk, and leaving the residue in a form that may be easily handled or stored. The filter-press, as usually constructed, consists of cast-iron plates or discs, with sunken drainage channels, and with a rim projecting from each face. When the plates are mounted in a suitable framework, these rims enclose cell-like spaces. The faces of the discs are covered with suitable filtering-cloth. The sludge is forced into these cells under heavy pressure, and the liquid passes through the filtering-cloth into the drainage grooves on the discs, whence it escapes, while the solids accumulate in the cells. When the liquid ceases to flow, the press is opened, and the compressed solids of the sludge, containing more or less moisture, are removed in the form of hard flat cakes. The sludge may be pumped into the press either by direct-acting pumps, or by the action of compressed air upon it, in strong air-tight iron tanks.[1] It is customary to add to the sludge, before pressing, some absorbent or solidifying material, such as ground slag, clay, plaster of Paris, or lime, to facilitate the drying and hardening of the cake. The cost of operating the filter-presses averages from 50 to 60 cents per ton of sludge-cake, which represents 5 tons of wet sludge, or the daily product of a population of about 3250.

[1] Mr. Wickstead, of Leicester, England, pumped the sludge directly into a stand-pipe, and used gravity pressure for driving out the surplus water, producing " a constant and steady pressure at little cost compared with the use of air."

The liquid squeezed out of the sludge in the filter-press is usually discharged into the outfall sewer, or into the precipitating tanks, to be re-treated. Where the quantity is small, there is practically no objection to this; but this liquid is exceedingly foul, and does not readily yield to further chemical treatment. The best disposition that can be made of it is to flow it over a tract of land.

Much has been said concerning the manurial value of the product of the filter-press, — "sludge-cake," — and it has been claimed that a profit on its manufacture could be realized. This is not to be hoped for, even under the most favorable circumstances. We shall do well if we succeed in realizing from it even a trifling amount against the cost of production. The safer course to adopt is to consider the sludge-cake as rubbish, to be got rid of in the cheapest possible way. It is obvious that its ingredients cannot be more valuable than the sum of the constituents of the untreated sewage and the chemicals used in treatment. Of course the latter can add nothing to the value beyond their own cost. As a matter of fact, a considerable proportion of them is lost, and the remainder exist in insoluble form — precipitate — which is generally worth considerably less than the reagent originally employed. Instead of adding to the value of the manure, certain precipitants — notably lime — actually diminish it, by dissolving some of the suspended matters, and favoring the production and escape of free ammonia.

The elements of value contained in the sewage are

N

nitrogen, which exists to the amount of about three ounces in each ton; potash, which averages half an ounce; and phosphoric acid, seven-eighths of an ounce per ton. If the entire amount of each of these ingredients could be arrested and stored in a form suitable for speedy assimilation by plants, sludge-cake would form a valuable manure. But this result cannot be obtained by any chemical process. Nitrogen, as free ammonia, and as alkaline nitrates and nitrites, cannot be precipitated by any reagent now known, and about 70 per cent of it escapes in the effluent in these forms. The remaining nitrogen exists in a form not readily assimilable, but as organic matter, which resolves into ammonia and nitric acid slowly, and only after further processes of change. The greater part of the phosphoric acid can be precipitated. This reduces the sludge-cake to about one-fourth of its theoretical value. Considering the cost of collection, treatment, and distribution, interest on investment, and the diminished value of the precipitating chemicals, it is clear that the manufacture of sewage sludge cannot be a profitable enterprise. Nevertheless, the process may help to solve a vexing problem of sewage disposal, and so prove an excellent investment.

Whether or not sludge-cake can be sold at all depends much more on local considerations than on theoretical value. Mr. Crookes, F.R.S., has rightly said:[1] "Of the value of a manure, chemistry can tell us little more than it can of the value of water. Just as a mere chemical analysis would utterly condemn water

[1] Quarterly Journal of Science, January, 1873.

containing Liebig's extract, infusion of tea or a glass
of bitter ale, as largely contaminated with nitrogenous
organic matter or albuminoid ammonia, so chemistry,
by taking a fictitious standard for manures, . . . gives
an arbitrary money value to a manure which is often
exceeded by the price it fetches in the market." The
market price often falls far short of the theoretical
value, and the selling of sewage manure may be disregarded in all our calculations.

What value the sludge has can be increased by drying
and pulverizing the cakes. At the Wrexham works,
in Wales, Lieutenant-Colonel Jones, after drying the
sludge until 80 per cent of its moisture has been
expelled, reinforces it by adding seven parts of raw
bone meal and one part of sulphate of ammonia to every
twelve parts of pulverized sludge. This, or some similar method of artificially fortifying the sludge, is frequently practised in England. It seems probable that
a judicious application of the principle might, at small
cost, in some cases, make all the difference between
rubbish which could only be disposed of at considerable
outlay, and a fertilizer which could be sold; that is to
say, an expense might thus be turned into a source of
income, but even this is doubtful. It has been suggested that if the sludge-cake cannot be sold as manure,
it may be utilized as fuel for the power plant used to
drive the mixing-machinery, pumps, etc. For this purpose the furnace should be so constructed as to ensure
complete combustion.

Having described the successive operations of chemical precipitation and the various methods for disposing of their products, we have now to consider the results to be obtained by their use, in the way of purification.

If a suitable process is employed, if chemicals are used in suitable quantities, and if the working of the plant is under careful, intelligent, and constant control, there is no doubt that practically all of the suspended matters in the sewage can be removed, and an effluent produced which is clear, colorless, and sparkling, and which will not froth when agitated.[1] A certain amount of dissolved matter, organic and inorganic, can also be withdrawn, including gelatine, mucus, albumen, and similar compounds, which constitute a considerable portion of urine, of blood, and of the soluble parts of the fæces, and which are likely, under favoring conditions, to undergo offensive decomposition. Nearly all of the phosphoric acid, which, as one of the constituents of urine, is always to be found in sewage, can be removed by precipitation. This is important; for although not in itself an organic impurity, its presence in the effluent would stimulate the growth of putrefactive as well as other germs. The remainder of the dissolved matter escapes with the effluent, and no known method of chemical treatment can prevent this. Increasing the quantity of chemicals beyond a certain point will only make matters worse; for an excess, especially if lime be used, will dissolve some of the suspended matters ordi-

---

[1] Some of the suspended matters, especially the fatty particles, are often difficult to deal with; still, if properly treated, they can be precipitated.

narily insoluble and make the effluent even more objectionable than raw sewage which has passed through filter-paper. A similar increase of the amount of impurity in the effluent may be caused by the decomposition of sludge which has accumulated in the bottom of the subsidence tanks, and which has not been removed at proper intervals. The amount of dissolved impurities that can actually be removed by precipitation ranges probably from 5 to 40 per cent, and the effluent water contains, under all ordinary conditions, a large quantity of putrescible matter, which, sooner or later, must be resolved into its elements. Where this effluent can be discharged into a moving body of water of sufficient volume, this destruction will be imperceptible and without offence; but where the diluting stream is small, the results may be far from satisfactory. "At Birmingham, Nottingham, Leicester, and elsewhere, the most perfect chemical processes have broken down, simply because there has not been sufficient water in the rivers receiving the effluents to complete its purification."[1]

Chemical precipitants have not merely the power of separating the inert impurities of sewage, but they may, to a great extent, remove its living organisms. These organisms exist, of course, as matters in suspension, and they are, like the other insoluble impurities, mechanically entangled in the coagulum which is formed, and are carried down with the sludge. In addition to this mechanical removal, certain reagents undoubtedly act as germicides. Alum has long been

[1] Sewage Disposal Works, W. Santo Crimp, p. 161.

used by the Chinese for purifying the water drawn for domestic purposes from their rivers, which serve alike as sources of water-supply and as sewers, and the French troops, when in Tonquin, adopted this means for improving the water, thereby almost entirely obviating the dysentery, which had caused much suffering among them. The investigations at the experiment station at Lawrence showed that the use of certain precipitants resulted in the removal of 98 per cent of the total number of bacteria existing in the sewage before treatment.

It can be stated in general terms — and it is admitted by the very champions of precipitation — that no chemical treatment, pure and simple, can produce an effluent free from organic impurity. It can clarify and deodorize sewage, and, to that extent, can purify it; but it cannot bring it to the condition of drinking-water, nor even to such a degree of purity as to warrant its discharge into a water-course which has not a large diluting flow.

The Committee of the British Association on the Treatment and Utilization of Sewage stated the following conclusions in their Report for 1873 (p. 449):—

"That the precipitation processes that it [the Committee] has examined are all incompetent, and necessarily so, to effect more than a separation of a small part of the valuable ingredients of sewage, and that only a partial purification is effected by them. Some of them may, however, be useful as methods of effecting a more rapid and complete separation of the sewage sludge."

Commenting on the above, Corfield and Parkes say:[1]

[1] Treatment and Utilization of Sewage, p. 351.

"All these precipitation processes do, then, *to a certain extent,* purify the sewage and prevent the pollution of rivers, chiefly by removing the suspended matters from the sewage; but they all leave a very large amount of putrescible matter in the effluent water, and at least all the ammonia contained in the sewage (sometimes they add to it); the greater part of the phosphoric acid is precipitated by some of them, while they increase the hardness of the river water, — a matter of great importance if the stream be a small one."

The Royal Commission on Metropolitan Sewage Discharge, in the Recommendations of its Second Report, said: —

"We are of the opinion that some process of deposition or precipitation should be used to separate the solid from the liquid portions of the sewage. . . . The liquid portion of the sewage, remaining after the precipitation of the solids, may, *as a preliminary and temporary measure,* be suffered to escape into the river. Its discharge should be rigorously limited to the period between high water and half ebb of each tide, and the top of the discharging orifice should be not less than six feet below low water of the lowest equinoctial spring tides. . . . We believe that the liquid so separated would not be sufficiently free from noxious matters as to allow of its being discharged at the present outfalls as a *permanent measure.* It would require further purification; and this, according to the present state of knowledge, can only be done effectually by its application to land."

A Commission appointed by the Municipal Authority of the city of Turin, to inquire into methods adopted for the disposal of the refuse of various towns in Europe, stated in its Report: —

" The chemical or precipitation methods for the treatment of sewage, which have been tried up to the present time, do not succeed in separating the manurial ingredients, are costly, clarify but do not purify the water, which, moreover, remains liable to undergo putrefaction afresh if the process is not followed by some method of oxidation."

Corfield and Parkes, in summarizing the results of their investigations of the subject, say: [1] —

" As to the utilization of sewage, we have shown the futility of all attempts at precipitation of its valuable constituents; in fact, 'it is hopeless,' as Dr. Hewlett says, 'by either one or any of these operations to render the effluent water anything else than sewage.' "

The Royal Commission on Metropolitan Sewage Discharge, in its Report for 1884, said: —

" No one denies that by any chemical precipitation the *suspended* matters may be almost entirely removed, or, in other words, the sewage may be practically clarified. It is proved that with well-devised, not too deep, and abundant tanks, so as to allow of complete subsidence (which may be well effected in a few hours), a clarified sewage may be prepared by precipitation, which will contain less than two or three grains of suspended solid matters per gallon. And as it is also

[1] Treatment and Utilization of Sewage (edition 1887), p. 498.

admitted that the suspended matters are the worst causes of pollution and nuisance, it follows that the clarification must effect a great improvement.

"It seems also to be the general opinion that the chemical processes in their best form will also have *some* effect in removing noxions matters in solution. It is difficult to say how much effect will be so produced. The amount has been differently estimated by different persons, and probably it may vary at different times, with different kinds of sewage, and under different modes of treatment, but it cannot be very large. All agree that a considerable amount of polluting matter must be left in the effluent. . . .

"Precipitating processes, though the same in principle as those of thirty years ago, have been greatly improved in detail, and when well worked, are effectual where the quantity of sewage is not very great, where the sewage can be promptly treated, and where there is a running stream into which the effluent can be discharged, in a proportion not exceeding 5 per cent of the supply of fresh water. . . .

"The effluent . . . must be brought as speedily as possible under an oxidizing influence, either by turning it into a stream containing sufficient oxygen to oxidize the organic materials, or by applying it to land, where it is also brought under powerful oxidizing influences.

"Should, however, the effluent be kept undiluted, or should it be turned into a stream in too large quantities for the free oxygen to deal with, the organisms or their spores, which have escaped in the effluent, multiply, and set up a renewed putrefaction. Such effluents, though apparently clear, become clouded, and a secondary deposit takes place in them. Bacterial fermentation of the cleanest of fluids is always attended by clouding and turbidity."

If a purer effluent is required than can be obtained by chemical precipitation alone, and if suitable land for irrigation or filtration is not available, a high degree of purification may be attained by precipitating the sewage, and afterwards passing the effluent through filter-beds of small area. The chemical treatment will remove practically all the suspended matters of the crude sewage and a certain amount of the dissolved impurities. Roughly speaking, one-fifth of the total impurities are in suspension, and four-fifths are in solution. Assuming that precipitation will remove 25 per cent of the dissolved matters, which is a safe average, two-fifths of the total impurity will be eliminated, including all of the suspended solids and slimy matters which tend to retard the process of filtration by mechanically obstructing the pores of the filter, and which necessitate for their destruction longer periods of rest than would be required for the mere aeration of the pores. The liquid to be purified will then be entirely free from sediment, and the filter will not become choked, but will always be in a condition to pass the liquid freely and to absorb air readily in periods of rest. It naturally follows that a filter which has only to deal with such a liquid will have much greater capacity for

work than one which receives crude sewage with its
grosser impurities, and a small area of land will suffice
for the purification of a precipitation effluent, when a
large area would be required for the cleansing of an
equal amount of raw sewage. The definite determination of the amount of ground actually needed for
completing the purification of the effluent from a
precipitation process will depend, of course, upon the
nature of the land available for the purpose. It is
probably safe to say that one acre will suffice for from
2000 to 10,000 people, according to local conditions.
In order that the standard purity of the effluent may be
maintained, and that due economy may be observed in
the use of reagents, it is necessary that frequent chemical and biological analyses of both the crude sewage
and the effluent should be made. The simple tests
for determining the treatment of the moment, described on pages 163 and 164, must be verified and
controlled by occasional more accurate determinations
of the result. No single analysis of a precipitation
effluent, and no group of analyses of samples taken on
different days but at the same hour, is sufficient proof
of the satisfactory working of a process. The composition and the strength of sewage varies continually, and,
in order that accurate conclusions as to the general
result may be reached, samples should be collected at
stated intervals, say hourly, during the whole twenty-
four hours, and these mixed in proportion to the volume
of flow at the time of the taking of each sample.
Analyses of the effluent should be made from samples

taken and mixed in a similar manner. Samples of sewage and effluent taken at about the same time cannot be compared, for the effluent will represent the sewage entering the tanks hours before the collection of the samples, and the sewage examined will not reach the outfall of the tanks for hours to come. When a special determination of the flow of any given time is desired, a float may be placed in the upper end of the tanks when the sample of sewage is taken, and the effluent may be sampled when the float reaches the point of discharge.

For ordinary purposes, the chemical analyses need be but partial. The chemical condition of the effluent, from a sanitary standpoint, will be clearly set forth by a determination of the free and albuminoid ammonia — the latter indicating with sufficient clearness the degree of purification attained — and of chlorine, which may, in a general way, be interpreted as showing the strength of the sewage treated. It is to be regretted that no standard system of stating the results of such analyses has been agreed upon, some chemists rendering their reports in "parts per 100,000," and others in "grains per gallon." A simple formula for translating from the former standard into the latter is

$$p = \text{parts per } 100,000.$$
$$x = \text{grains per gallon.}$$
$$x = \frac{p \times 3.5}{6}.$$

In estimating the cost of treating sewage chemically, the following must be considered: (1) The volume of sewage to be dealt with. Large quantities can be treated more easily and more cheaply than small quantities, and the cost of laboratory experiments is practically of no value in determining the expense of operating works on a large scale.[1] (2) The quality of the sewage, and the possible presence in it of unusual elements, as of certain manufacturing wastes, requiring special treatment. (3) The location of the proposed works, as affecting the handling of sewage, effluent, and sludge. Unless sufficient fall can be obtained for gravity flow, the cost of pumping must be included. (4) The local cost of the chemicals to be used, and the amount necessary to produce the results required by the adopted standard of purity. (5) The bulk of sludge produced, its value,— if it has any,— or the cost of disposing of it if worthless.

The actual operating expenses will include the cost of chemicals, of power, of labor, and of repairs and renewal of plant, etc. To these must be added interest on the cost of original construction. The cost of chemicals, which depends on the quantity and quality of sewage, and the method and degree of purification adopted, will vary from fifteen to fifty cents per capita yearly. The cost of power must be calculated in each

---

[1] In their Report to the Metropolitan Board of Works, concerning chemical treatment of the sewage of London, Dr. Frankland and Dr. Hoffman say: " We beg to express our opinion, based upon the experience acquired during this investigation, that the disinfection of vast volumes of sewage can be more easily accomplished than is generally believed, and than we ourselves anticipated at the commencement of our inquiry."

case independently, as the conditions vary widely. Elevators, grinding mills, sifters, and mixers for the chemicals; agitators for mixing them with the sewage; sludge-pumps, filter-presses, etc.; any or all of these may be needed according to the special circumstances of the case under consideration. The amount of labor required will also be a variable factor in the problem, and its cost will be determined largely by the manipulation necessary and by the method of disposing of the sludge. From 5 to 7 per cent of the cost of construction should be allowed for repairs, renewal of plant, etc. The total annual cost of chemical precipitation per capita will, under ordinary conditions, range from 45 cents to $1.10. As has already been stated, no return from the sale of sludge can be counted on. Any income that may be realized from this source will, of course, be a welcome contribution towards the expense of maintenance, but we must regard the sludge as a thing to get rid of, and must be prepared to pay for its removal.

In 1889, Mr. Allen Hazen, Chemist at the Lawrence Experiment Station of the Massachusetts Board of Health, made a very thorough experimental study of the purification of sewage by chemical treatment, and the results of his investigations, which are of great value, are set forth at length in the Report of the Board for 1890 (pp. 737 to 791).

The precipitants tested in these experiments were those most commonly used in precipitation works, — *i.e.* lime, alum, copperas, and sulphate of iron, — and they were employed separately and in various combinations.

The object of the experiments was to determine (1) The method of using each precipitant that produces the best result. (2) The amount of each precipitant that yields the best result with sewage of a fixed composition. (3) The chemicals which produce the best effects in combination, and the proportions which give the maximum efficiency. (4) The comparative effects of the different precipitants, when applied to sewage of a fixed composition in amounts representing equal money values, each precipitant being used in the particular method that develops its maximum efficiency. (5) The maximum amount of purification that can be obtained under the most favorable conditions. In all of the experiments the composition of the original sewage and of the effluent was determined, but the sludge was neglected altogether.

The first experiments were made in a large settling tank, holding about 700 gallons. The chemicals in solution were added to the sewage in the trough leading to the tank. When the tank was full, the flow was stopped and the solids allowed to settle through the motionless liquid; but in one experiment the tank was used for continuous precipitation, the effluent passing out at the lower end as fast as the sewage entered at the upper. These conditions resembled, as closely as was possible under the circumstances, those of a typical precipitation plant on a large scale; but the method was found unsuitable for the purpose of the experiments, —i.e. the determination of the comparative results obtained by the use of various chemicals, — in that, owing to the ever-changing composition of the sewage, it was impossible to make an exact comparison of the effluent with the original sewage. It also precluded the possibility of repeating any given experiment under the exact conditions which first obtained. To avoid these difficulties, a tank was filled with sewage and the inflow stopped. The collected sewage was then thoroughly stirred to secure a uniformity of composition, and a number of barrels were filled with it. One barrel was left to settle by simple sedimentation, for purposes of comparison, and the others were treated with the various chemicals to be tested. The capacity of the barrels being known, the proportion of chemicals added to each was easily computed. In this way comparable results were obtained. Each barrel was allowed to settle one hour, which probably equals two or three hours' subsidence in a tank six feet deep. The sample of effluent for analysis was then drawn from a tap about ten inches from the bottom of the barrel, and the determination of organic matter was made within three hours of its collection.

The results of Mr. Hazen's experiments show that,

With the ordinary sewage of Lawrence, "the most satisfactory results were those obtained with sulphate of iron. The second in order were with copperas and lime; the third were with lime; and the least satisfactory were with alum. . . .

"There is a definite quantity of lime that will give the best results with any given sewage. This quantity depends upon the character of the sewage, and is the quantity that is necessary to neutralize the carbonic acid. This quantity is the most economical to use, as a larger quantity would produce no better result, and a smaller quantity would produce a result poorer in proportion to its cost. With the ordinary sewage at Lawrence, this quantity of lime is found to be about 1800 pounds for 1,000,000 gallons of sewage; and, at the wholesale price there of $9 a ton, the annual cost of lime per inhabitant, upon the assumption that there are 100 gallons of sewage daily for each, would be 30 cents.

" With the other precipitants used,— viz. alum, copperas and lime, and ferric sulphate,— the purification, with the same cost of precipitants, was not quite so good with alum as with lime, and was a little better with copperas and lime, and with ferric sulphate.

"At greater cost, these other precipitants caused more impurities to be removed; but the increased amount so removed was not proportionately so great as the increased cost. . . .

" The lime process has little to recommend it. Owing to the large amount of lime-water required, and the difficulty of accurately adjusting the lime to the sewage, very close supervision would be required to obtain a good result, and, even then, the result is inferior to that obtained in other ways.

"Precipitation by copperas is also somewhat complicated, owing to the necessity of getting the right amount of lime mixed with the sewage before adding the copperas. When this is done, a good result is obtained. The amount of iron left in the effluent is much greater

o

than with ferric sulphate, owing to the greater solubility of ferrous hydroxide.

"Ferric sulphate and alum have the advantage over both lime and copperas, that their addition in concentrated solution can be accurately controlled, and the success of the operation does not depend upon the accurate adjustment of lime or any chemical to the sewage.

"The results with ferric sulphate have been, on the whole, more satisfactory than those with alum. This seems to be due in part to the greater rapidity with which precipitation takes place, and in part to the greater weight of the precipitate. It is probable, from the greater ease with which ferric sulphate is precipitated, that it would give a good result with a sewage that was not sufficiently alkaline to precipitate alum at once."

Sewage is usually sufficiently alkaline to decompose either alum or sulphate of iron without the addition of lime, which, under such conditions, is practically without effect.

"To make as definite a comparison as practicable, on two occasions portions of sewage well mixed, from the same tank, were treated with each of the precipitants, in quantities that would, at present, cost 30 cents per year for treating 100 gallons of sewage daily. The average results are placed in the following table, together with results obtained by allowing the same sewage to settle one hour without any precipitant, and those obtained in the laboratory by filtering the same sewage through filter-paper, and the general results obtained by intermittent filtration through four or five feet of sand: —

CONDITIONS OF TREATMENT.	Percentage of the Albuminoid Ammonia of the Sewage that was removed.
Settled one hour, —	
With 1800 pounds of lime per 1,000,000 gallons . .	52
With 650 pounds of alum per 1,000,000 gallons . . .	51
With 1000 pounds of copperas and 700 pounds of lime per 1,000,000 gallons	57
With 270 pounds of ferric oxide in the form of ferric sulphate per 1,000,000 gallons	59
With no precipitant	21
Filtered through paper	39
Filtered intermittently through four or five feet of sand	98

"When the amounts of alum and of ferric oxide were increased one-third, making their yearly cost 40 cents for 100 gallons daily, the amount of albuminoid ammonia removed was increased nearly 10 per cent, being, for alum, 56 per cent, and for ferric sulphate, 64 per cent.

"An average of all of the experiments shows that, with a cost of 30 cents as above, copperas and lime mixed in the proportions above given, and ferric sulphate, remove about 58 per cent of the albuminoid ammonia; that lime removes about 55 per cent, and alum removes about 51 per cent.

"At higher costs, copperas with lime, and ferric sulphate, remove increasing quantities up to about 66 per cent — when the cost is 45 cents — with indications that there would be little if any increase above this percentage at still higher costs. Alum removes increasing percentages at costs above 30 cents, but the results are five or six per cent below those with ferric sulphate. . . .

"It is quite possible that the same process would not give equally good results upon all kinds of sewage. Special sewages may require special treatment. For this reason, and also on account of changes in the prices of the several chemicals, it is impossible to say that one precipitant is universally better than another. . . .

"The general result with these precipitants is, that from one-half to two-thirds of the organic impurities of the sewage, as indicated by the albuminoid ammonia, may be removed by chemical precipitation, while 98 per cent may be removed by intermittent filtration; or, there remains in the effluent, after chemical precipitation, from sixteen to twenty-four times as much, or an average of twenty times as much organic impurity as in the effluent from intermittent filtration."

All organic matter is not equally subject to bacterial invasion, and there is reason for believing that the organic matter escaping in the effluent from a precipitation process is not so susceptible to the attacks of putrefactive germs as the same amount of organic matter in untreated sewage. This subject requires further investigation.

"The number of bacteria remaining in the effluent, after chemical precipitation, averages, from these experiments, about five per cent of those in the sewage; while the number found in the effluent from two of the filters . . . averages five one-hundredths of one per cent of those in the sewage, and the number found in the effluent from three of the filters averages but two-thousandths of one per cent of the number in the sewage, and none of these are believed to come from the sewage.

"The best results that we have obtained by chemical precipitation — and we know of no others that are so good — leave as much as one-third of the nitrogenous organic matter of the sewage in the effluent; this is an abundant food-supply for the unlimited growth of the large number of bacteria that remain. The number is called large, because 5 per cent of 700,000, or 35,000, in a thimbleful, is a large number; and, if any of these are disease-producing germs, there would be no safety in turning such a liquid into a drinking-water stream; and, whether it would be admissible to turn a liquid containing from one-third to one-half as much nitrogenous organic matter as sewage, with abundant bacteria, into any other stream, would depend upon nearly the same conditions that would attend discharging a less amount of sewage into the same stream. There would, however, be this difference, and it is an important one, — the objectionable appearance would have been removed, and would not come again, unless, collecting in pools or in eddies or on flats, or rising to the surface on a liquid having greater specific gravity, putrefaction of the remaining organic matter should follow.

"The remaining organic matter would probably not putrefy as readily as the original sewage diluted to the same extent; but that it is not so stable a compound that it will not readily decompose under favorable conditions, is shown by the fact that five-sixths of it may be nitrified while moving slowly for one day over gravel stones in intermittent filtration.

"Such an effluent as may be obtained from chemical precipitation of sewage, turned into a large and rapidly flowing stream, or into a tidal current that would soon take it to sea, would be disposed of without making a nuisance, when crude sewage might be very objectionable. Under such circumstances, and there may be others where the conditions for intermittent filtration are unfavorable, the partial purification of sewage by chemical precipitation may be the best practicable way to avoid a nuisance. But the incompleteness of the purification and the cost of thirty cents per inhabitant yearly for chemicals, together with the additional expense of manipulation and disposing of the sludge, will be likely to confine the application of chemical precipitation in the purification of sewage to narrow limits."

CHAPTER XV.

MESSRS. Rafter and Baker, in their elaborate treatise,[1] devote a carefully prepared chapter, "Legal Aspects of the Case," to a discussion of the law relating to the fouling of streams. In this it seems to be established that the ownership of land over which (and past which?) a water-course flows carries with it an ownership of the water while so passing, with the right to make any use of it that may be desired for business or pleasure; but the obligation is imposed on the owner to deliver the usual flow to his neighbors below in substantially unimpaired condition and quantity.

A town through or past which a stream flows has the same rights, but is restrained by the same obligations. It may use the water for the service of its people, but it may not materially diminish its volume, nor so pollute it as to make it materially less useful or agreeable for those who live further down the stream. The courts have held that these "are entitled to the use and enjoyment of the water, and to have the same flow in its natural and accustomed course, without obstruction, diversion, or corruption." This is a natural right,

[1] Sewage Disposal in the United States, 1894.

"which is not lost until an adverse easement has been acquired." To pollute a public stream is to maintain a common nuisance, and this is a crime. It is no defence to say that the pollution is of long standing, "for the continuance of the nuisance is itself an offence against the law. . . . No length of time can justify a public nuisance, although it may furnish an answer to an action for private injury," nor is it a defence that there are other sources of pollution: "If the defendants have contributed to the pollution, they are guilty."

The fact that an enforcement of the law against manufacturers who pollute a stream would injure large business interests is no reason for not enforcing it. "All persons engaged in business are bound to conduct that business in subordination to the law, and in such a manner as not to injure the public."

There are ordinary or natural uses, and extraordinary or artificial uses, — such as domestic use and the watering of stock on the one hand, and the sprinkling of streets on the other. For the natural uses, all the water of a small stream may be consumed by a single owner living on its banks, but he may not so consume it for artificial uses. A city may not so divert the supply of a stream by its water-works as to interfere with the rights of others to a sufficient supply for their use and enjoyment.

All artificial pollutions, "rendering water unfit for domestic, culinary, or mining purposes, for cattle to drink, fish to live in, or for use in manufacturing," are actionable. "So long as the reasonable use of the common property does no injury to the rights of others who are entitled to a like reasonable use, no action lies; but an unreasonable use is an actionable injury." It is for juries to decide what is reasonable under the circumstances of a given case.

Riparian owners can abrogate their rights and give up the stream to artificial uses, as for the turning of mill-wheels, and long custom may be taken to constitute such abrogation. Special easements may also be created "by deed, devise, or record." Under the common law, a right which might have been created by grant may be acquired by long-continued and unopposed possession; it becomes a right by prescription. The period required to acquire such prescriptive rights varies in different States, generally from fifteen to twenty years. It has been held that such prescription may attach to long-continued and undisturbed pollution of a stream. Lord Ellenborough said that, "although the stream may be diminished in quantity, or corrupted in quality, yet, if the occupation of the party so taking it and using it have existed for so long a time as may raise the presumption of a grant, the other party whose land is below must take the stream subject to such adverse right."

Rafter says, "It is believed, however, that in the present understanding of things, such a view is bad law, and an attempt will be made to make good that proposition." He argues the question with care, and advances the idea that it is inconsistent for human beings to pollute that which for their own use should remain unpolluted, and that "when we understand, as we now do, the serious effects of such pollution, the custom of turning sewage into any stream, which either is, or in the future may be, the source of public water-supply, is shown to be so utterly bad as to be worthy only of immediate abatement, even although the custom has been maintained from time immemorial."

The conclusion from the argument and from the cases cited would seem to be that the right to pollute a stream may be acquired by prescription,—as by twenty years' continued and undisturbed use,—but that the right is a limited one, so that "one proprietor cannot acquire the right by prescription to pollute the stream to a greater extent than it was polluted at the commencement of the twenty years." It would result from this, for example, that a town that had acquired a right to discharge a limited amount of filth into a stream through an open drain, would not have the right to discharge a greater amount of filth through a system of sewers.

Our courts have held that, while a prescriptive right may hold against private interests, no public nuisance can be maintained by prescription, however long its continuance may have been.

CHAPTER XVI.

THERE are hundreds of hotels, prisons, asylums, hospitals, colleges, etc., scattered over the whole country, which experience an increasing need for improvement of their means of sewage disposal. There is usually sufficient land available for the adoption of some form of irrigation or filtration, and the methods and principles set forth in the preceding chapters are applicable here.

Although such methods of disposal have not as yet been generally adopted in these cases, they have had sufficient use to demonstrate their entire efficiency and their superiority wherever the surrounding conditions are suitable. They offer the most promising solution of the very difficult problem which the gathering of a large number of persons in buildings beyond the reach of sewers always presents.

The first works of the kind were those of the Massachusetts State Hospital for the Insane, at Worcester, designed by Buttrick & Wheeler, civil engineers, in 1876. The ordinary population of the Hospital from that time until the present has been about 600, and the operation of the system has always been successful. Most of the sewage is delivered at a considerable distance from the building and on gravelly land, but the arrangements include the necessary pipes and pumping facilities for irrigating the ground immediately in front of the administration building, where sewage, in its fresh state, is spread directly over the surface of a large, sloping lawn immediately adjoining the main-entrance drive, and within a very short distance of the windows of rooms occupied by both officers and patients. The disposal on both tracts has been without offence and entirely effective. The sentimental objection to such delivery of sewage on pleasure-ground almost adjoining the building would seem to be very great, but there is only the sentimental objection. Practically, all that is necessary is to screen out from the sewage paper and other matters which would show on the surface of the lawn. What remains to be delivered contains often less, and rarely more, than one part in five hundred of anything but pure water. Such a very dilute mixture may be treated practically as ordinary brook-water might be treated, up to the time when it enters into a state of putrefaction, from which, and not from foreign substances as such, the real source of offence arises.

This point cannot be too strongly insisted upon. It constitutes the key-note of all successful sewage disposal. Before putrefaction begins sewage has no offensive odor; after it begins to putrefy, it rapidly becomes both offensive and dangerous. Therefore, no system of surface disposal can be entirely satisfactory from which the element of putrefaction is not eliminated. Even such small receptacles for retaining sewage as ordinary kitchen grease-traps, large catch-basins, cesspools, and all possible seats of putrefaction should be regarded as objectionable in any system of house drainage or sewerage of which the outflow is to be treated by surface irrigation or filtration, or, indeed, by chemical precipitation either.

In the competition for the location of the Eastern Insane Asylum of Pennsylvania, the city of Norristown made a very strong and successful argument, based on the exceptionally good drainage facilities offered by a small stream, Stony Creek, flowing by the proposed site and through the city to the Schuylkill River. Soon after the occupation of the Asylum, a loud outcry was raised against the intolerable fouling of Stony Creek, and the writer was engaged to devise means for artificial removal. The details of the system adopted have already been extensively illustrated and described.

The sewage is received in an open reservoir 40 feet square and 7 feet deep. When this reservoir becomes filled, it is discharged by the operation of an automatic siphon. Toward the end of the discharge, the walls of the reservoir are washed down by an automatic sprinkling pipe, and its floor is frequently washed with a hose. The outflow, amounting to about 75,000 gallons, is delivered through an 8-inch pipe which passes through a valley under a creek, and rises about 15 feet to a carrier running along the far side of the irrigation field, from which it is delivered alternately over each of three sections into which the field is divided, several discharges being applied in succession to each section. The tank discharges three or four times in twenty-four hours. This system was first put in operation June, 1885. In June, 1887, the President of the Board of Trustees wrote: —

"Since your system was put into operation at the hospital, every doubt has given place to a conviction that the problem of sewerage for public institutions has been successfully solved. It has been a success from the first, and the Board of Trustees and the public are entirely satisfied with the results."

Up to the last accounts, 1892, the system was performing its work in a most satisfactory manner.

In 1889, the writer was engaged to provide for the disposal of the sewage of the Insane Asylum near London, Canada. The work involved the reconstruction of much of the plumbing work of the establishment, and the entire abandonment of the old system of outside drainage, which had been through combined sewers taking the roof-water and foul sewage of all the buildings to the head-waters of a brook which passes through the city. To satisfy complaints that had been made as to the fouling of the brook, a rude sort of settling-basin had been established to hold back solid matters. This failed to correct the difficulty. In arranging the new work, this old system was left for the removal of roof and ground water only.

All of the buildings have their soil-pipes, kitchen-drains, laundry-wastes, etc., connected with a system of

6-inch sewers having tightly cemented joints and with
carefully graded inclinations, leading to a central tank.
Each branch of these sewers is provided at its head with
an automatic flush-tank, preventing deposits in the
sewers, and so preventing putrefaction. The central
tank is 70 feet by 40 feet; its walls are 16 inches thick;
its bottom is of concrete. It is covered by three longitudinal arches, 12.66 feet span, 12 inches thick. These
arches rest on two longitudinal walls with arched
openings. The floor of the tank is suitably graded,
varying between elevation 31.9 and 32.3 respectively.
Each section has a longitudinal drainage-gutter, with
its upper end at 32.22, and its lower end at 31.98,
31.94, 31.90, with a cross gutter leading to a sump four
feet in diameter with its bottom at grade 30°. The
bottom of this sump is hemispherical, and the suction
of the pump is centrally located, having six inches
space between its mouth, which is bell-shaped, and the
bottom. The elevation of the ground at this point is
47.5, making the surface of the floor of the tank about
15 feet below the surface. There are three man-holes
at each end of the tank, with covers at the surface of
the ground. At the receiving end of the tank, at the
head of the central chamber, is a screening-chamber,
reaching to the surface of the ground and with its
bottom at elevation 34.4. The opening from this
chamber into the tank is 8.33 feet wide, and it is
provided with a screen carried in slots in the side walls
4.5 feet high in the centre. The screen is made of
wrought-iron and galvanized. The vertical bars are of

half-inch round iron, and the openings between them are one inch wide. The top of this screening-chamber is covered at the surface of the ground with a hinged wooden cover.

The tank is intended to be filled to a depth of 5 feet, or to the spring of the arches, to which height it has a capacity of a little more than 100,000 gallons. It is located near the main fire-stack of the establishment, adjoining which a basement pump house has been constructed, having its floor at such an elevation that when the tank is filled to the spring line, the sewage rises through the suction-pipe into the wheel of the centrifugal pump, filling it above its axis, so that when the pump is started, the driving out of this water is sufficient to exhaust the pump of its air and establish a full suction. The pump is an 8-inch Webber centrifugal, with Westinghouse engines working directly on its main shaft. The suction is 10 inches in diameter, and the force-main 8 inches. As a precaution against a possible failure of the pump, a 6-inch overflow is provided at the top of the tank, which delivers into the main brick sewer of the old drainage system, passing near by.

As the tank is underground, artificial ventilation became necessary. This is accomplished through the man-holes. Those at the end farthest from the pump are covered with open gratings. Those at the end nearest to the pump have tight covers. A 10-inch ventilator is taken from the side of each tight man-hole, and is carried to a junction chamber with a 15-inch ventilation-pipe connected with the chimney. There

is thus maintained a constant current down through the
perforated covers, through the whole length of each
section of the tank, and out by the man-hole shafts,
which are connected with the chimney.

The sewage is delivered to the tank by three 6-inch
lines entering the screening-chamber built out from its
end. Two of these sewers deliver at an elevation
higher than the top of the tank; the other delivers at
the level of the spring line, bringing drainage from
cottages on somewhat lower land than that occupied by
the main buildings. The screen is movable, so that it
may be taken out for cleansing when occasion requires.
It is sufficiently open to admit all but the larger objects
brought in by the sewage to the tank and pump. It is
a great advantage of the centrifugal pump that it so
beats and thrashes the sewage as to reduce nearly all of
its solids to atoms. The force-main is an 8-inch spiral,
riveted pipe leading from the centrifugal pump to the
receiving-well at the absorption field, 1550 feet distant.
This 8-inch force-main has a continuous rise from the
pump to the receiving-well, and the pump has no valve,
so that whenever pumping ceases the contents of the
receiving-well and force-main flow back into the tank,
thus obviating the possibility of freezing.

The field devoted to sewage disposal contains about
thirty acres, level at its upper end and sloping gradually
for the remainder of the distance to an open water-
course.

The level portion of the field, about 5 acres, is laid off
in ditches separated by beds. The ditches are in pairs,

returning on each other.   Under the bank between each pair is a deep underdrain of agricultural tile.   The soil is very light and sandy over the greater part of this tract, and thus far the removal of the purified sewage has been by underground soakage, even the descending effluent, which enters the tiles in the heavier portions in the ground, leaking out through their joints before the outlet is reached.   The main outlet from the receiving-well (18-inch channel-pipe) runs across the end of the field, and has a branch to each of the ditches.   It has a fall of 1 to 500.   At its lower end it delivers into a distributing ditch, which, at its lower end, is connected by a carrier ditch with two other distributing-ditches farther down the slope.   The main outlet from the receiving-well is made of vitrified 18-inch pipe, split in halves.   These half-pipes are connected by concrete channels with the ends of the parallel ditches.   There are gate-slots between the half-pipes and the concrete branch pieces, furnished with movable gates.   By placing and removing these gates, the flow of sewage can be directed at pleasure into all or any of the pairs of parallel ditches.   The connection between the concrete branch pieces and the ditches is made with two lengths of vitrified pipe (four feet).   As the bottoms of the ditches are all in the same plane, and as the main outlet from the well has a fall, there is a drop of varying height from the half-pipes into the ditches.   At this point, and even where the drop runs out at the lower ditch, the bottom of the ditch is roughly but strongly paved, to check the flow and prevent the cutting of the bottom at that point. The absorption ditches are eight feet wide at the top, two feet wide at the bottom, and one and a half feet deep. They are separated by beds ten feet wide at the surface. This level area, with its settling-ditches, may be used for intermittent downward filtration, and as the total capacity of the ditches is equal to twice the capacity of the tank, even were there no immediate filtration, the area is worked in two or three sections alternately. Two or four of these ditches at the lower side of the field may be used, if found necessary, as settling-ditches to deposit heavy matters before delivering the liquid over the surface of the irrigation-tract below. It has been found, however, that the churning of the sewage given by the pump makes such treatment unnecessary.

The capacity of the ditches of the level tract has proved in practice to be so great as to require little use of the distributing-ditches in the field below; they are, however, useful for the irrigation of crops. These ditches have a fall of 1 to 500 and are used in the following manner: —

If the flow through a distributing-ditch is arrested at any point, as it may be by striking a wrought-iron gate into the earth, making a dam across it to above the top, the sewage will overflow for a greater or less distance above the dam, according to the volume of the current. If the dam is placed first at the lower end of the upper distributing-ditch this may overflow, for example, 200 feet above the dam. When the ground to be reached by this overflow has received a sufficient supply of sewage, the dam is placed about 200 feet higher up stream, and the overflow carried over the next section, and then, in like manner, over a third. Should the ground between the two ditches not be able to absorb all the sewage discharged upon it, the overflow will be caught by the lower one, and if its quantity is sufficient can have its distribution regulated by the placing of a dam there, as above described.

The main outlet from the well is 400 feet long; the settling-ditches have an aggregate length of 3600 feet; the carrier and distributing-ditches have an aggregate length of 3100 feet; and the tile drains aggregate 6600 feet. The outlet to the under drainage (6-inch tile) is 6 feet deep at the end of the upper bed, 6½ feet deep at the lower bed. The lateral drains are 4-inch tile for the lower half, and 3-inch tile for the upper half. The upper ends of these laterals are 4 feet below the surface of the beds, and they are carried on a true grade to the 6-inch outlet pipe. This system was put in operation in July, 1889, and has worked ever since in the most satisfactory manner. The Superintendent of the Asylum wrote as follows (January 2, 1891): —

"Our sewage-disposal works, put in in the early part of 1889, have been a complete success; neither snow, frost, nor anything else has so far interfered, materially, for a single day with their operation, and I consider that the problem of sewage disposal is solved here for all time."

In a recent letter (January 15, 1894) the same official says that "the sewage field has been and is an absolute success. Over and above complete and perfect sewage disposal, we last year raised a crop of the value of $850 on the four acres of which the field is composed, *i.e.* on the two acres of beds between the depressions."

A system of sewage disposal was constructed under the writer's direction, in 1893, for the Maryland Hospital for the Insane, at Catonsville. In the official report of the superintendent, it was stated that "the absorption of water and destruction of organic matter is complete. No offensive odor is given off from the field, or indeed from any part of the system, and the outflowing water, after filtration through the land, has been shown by chemical analysis to be as pure as ordinary drinking-water."

This disposal field is close to the hospital buildings, and borders a carriage drive.

CHAPTER XVII.

IN providing for the better drainage of country houses, and of houses in villages which are without sewers, — houses which especially need such provision, — the principles hereinbefore set forth are perfectly applicable. The safe and inoffensive disposal of the sewage of such houses has always presented difficulties.

Until twenty years ago, the "out-of-sight-out-of-mind" system — in other words, the use of cesspools — was nearly universal. The radical objections to the cesspool had already begun to be recognized, and they are now known to be absolutely condemnatory. Cesspools are always objectionable, and there are not many cases where they are at all admissible.

Soon after the introduction of the earth-closet, its inventor, Mr. Moule, devised the method of discharging sewage into shallow, underground drains, and this was soon improved and brought to something like a system by Mr. Rogers Field, who collected into a flush-tank the dribbling stream of sewage from the house, and caused an intermittent discharge of considerable volume, and at considerable intervals, into the tiles. This system was somewhat improved and introduced into this country by the writer, and has been brought to such a state of efficiency as to come into quite general use. It is often called "Waring's system," which is a misnomer; it is, in its origin, much more Mr. Field's system. Among other large works of this kind may be mentioned those planned by the writer for the disposal of the sewage of the Women's Prison at Sherborn, Mass., and of the village of Lenox, Mass. Somewhat later Mr. Croes used the same system at the Lawrenceville School in New Jersey, discharging his tank by intermittent pumping, instead of the intermittent action of an automatic siphon. For domestic use in connection with suburban houses, the system has been very widely adopted.

In 1884, the writer used a flush-tank for the accumulation of the sewage of a large country house near Baltimore, delivering its discharge with satisfactory results directly on to the surface of the ground, instead of into sub-surface absorption-tiles.

When sewage is delivered into porous absorption-drains laid in the surface soil, it soaks away into ground, which is sufficiently penetrated by air to facilitate bacterial growth. When sewage is delivered over the surface of the ground the process is substantially the same, but exposure to the air is more complete, and oxidation is, within suitable range of temperature, proportionately active. In all work of sewage disposal, the conditions essential to bacterial oxidation must be kept constantly in mind, and as much as possible favored. The process being in all cases the same, the skill and judgment of the engineer will be applied to regulating the work in such a manner as to secure its best development and most favorable action.

Aside from the efficiency of the ultimate disposal, it is of the greatest importance, when the works are established in the immediate vicinity of the buildings in which the sewage is produced, that all parts of the process should be free from offence to the eye and to the nostril. The system of sub-surface irrigation, with the use of a double-chambered flush-tank, the first chamber holding back solid matters and scum, and the second chamber accumulating and intermittently discharging the liquid portions, has always had the drawback, more theoretical than practical, that the first chamber had, necessarily, some of the worst characteristics of a cesspool; for, while the free movement of liquid through it prevents a high degree of foul putrefaction of any of its liquid portions, and indeed carries off the gases of the putrefying sediment, the scum with which the contents of the tank are always covered is in a constant state of decomposition, and is constantly producing foul and objectionable vapors. No way has yet been discovered in which this foul deposit-chamber can be dispensed with in the case of sub-surface disposal with the absorption-drains generally used. As, however, it is absolutely tight and cannot foul the ground in which it stands, and as its gases are so trapped off from the house that they can only escape imperceptibly to the outer air, this drawback is reduced to a minimum, and it has not been shown in practice to be a detriment.

Although it was not the intention to enter in this volume so far into the practical technicalities of work as to use any considerable number of diagrams in illustration, it is thought that, as the system under discussion, though largely used in a few Eastern neighborhoods, is still little known over the country generally, it will be proper to make an exception here, and to reproduce, from a descriptive article, published in the *American Architect* for March, 1892, the diagrams accompanying it.

Fig. 5

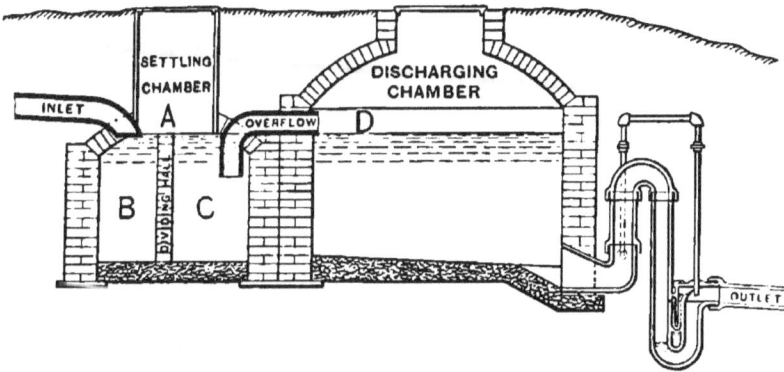

Figure 5 shows the construction of the double-chamber tank. The settling-chamber *A* is a small, round cistern with a wide throat — not less than eighteen inches diameter — to facilitate the removal of its scum and deposit. It receives sewage from a pipe turned down through the dome and barely trapped against the return of air — if deeply trapped, grease accumulates and obstructs the drain; with this slight trap, the flow from an ordinary house suffices to keep it free. It overflows through a deeply-trapped pipe into the discharging-chamber *D*. It is divided by a wall into two chambers, the top of the wall being just at the overflow line. The compartment *B*, on the inlet side, has its water considerably agitated by the inflow. Before the dividing-wall was adopted, this agitation was communicated to the contents of the whole chamber, and flocculent matters, which would settle to the bottom or float to the top in still water, were carried over by the current into the discharging-chamber. This agitation is now confined to the compartment *B*, from which the liquid portion flows to the compartment *C* in a thin sheet over the top of the wall, in such a manner as not to disturb the contents of this second compartment, allowing flocculent solids to settle quietly to its bottom. Under some circumstances, perhaps due to a higher temperature in the sewage, and this to its larger amount, the decomposition of the sediment and of the scum is sufficiently active to prevent accumulation to an injurious amount. In such cases, the settling-chamber need never be cleaned. This is not to be depended upon without occasional inspection. In the majority of cases it is necessary every few months to bail out the chamber and get rid of its accumulations, which should be buried or dug into the ground at once.

The liquid overflow from the settling-chamber *A* to the discharging-chamber *D* represents practically the full amount of sewage brought down by the drain. The discharging-chamber should be made large enough to hold the product of at least twelve hours; there is no objection to its retaining twenty-four hours' supply,—a longer retention would lead to too much putrefaction. This chamber is furnished with an automatic siphon : the one shown is what is known as the Rhoads-Williams siphon.   Its details are shown in Figure 6.

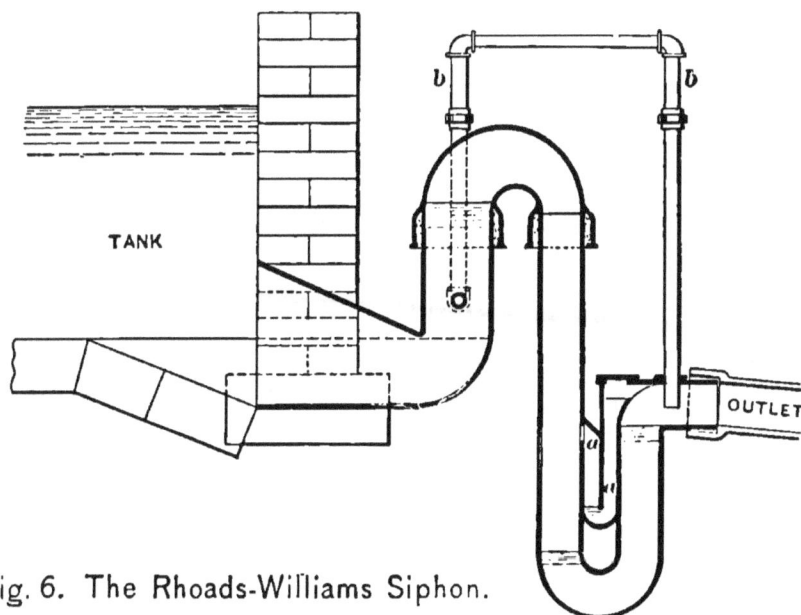

Fig. 6. The Rhoads-Williams Siphon.

It depends for its action on the sudden releasing of compressed air contained between the inflow from the tank and the deep trap near the outlet. When the pressure is sufficient to force the water in the blow-off trap *a, a* to the bottom of this trap, the air-pressure is released, and the head of water, which it had held in the tank, forces a full flow into the siphon and brings it rapidly into action.   Air is introduced for the breaking of the siphon after the main flow has ceased by the admission of air from the drain through the pipe b, b. These siphons are sold by flush-tank dealers.

The siphon is located entirely outside of the tank. This obviates the serious fouling of the siphon itself, which has always been a source of difficulty when it was placed within the tank. Its opening into the tank is funnel-shaped so as to take the flow rapidly, and to make sure that there will be no obstruction from such minor solid matters as the sewage may contain. It is not ordinarily found necessary to clean out the discharging-chamber, matters which would otherwise accumulate within it being held back in the settling-chamber.

The outflow is a slightly putrid sewage containing more or less fine flocculent matter, not enough to interfere with the proper action of 2-inch absorption-drains. These absorption-drains may be placed at a greater or less distance from the tank as the tank may be at a greater or less distance from the house. They are made of ordinary round 2-inch tile in one-foot lengths, laid in earthenware gutters, their joints being open about one-fourth inch, and being protected against the entrance of earth by the loose-fitting cap laid on the top. The gutters and caps are of larger radius than the outside of the tile, so that practically the whole joint is available for the escape of sewage into the ground. The surface of the gutter on which the tile is laid should be ten inches below the finished surface of the ground. In a reasonably porous surface-loam, it will suffice to have one foot of tile for each gallon of the contents of the discharging-chamber.   The tile, caps, and gutters are shown in Figure 21.

If the soil is heavier, the length must be increased. An impervious clay is not well suited, under any circumstances, for this use, but where nothing else is available there should be at least three feet of tile per gallon.   The tile may be one continuous line, or a number of shorter lines, connected with the 4-inch main leading from the flush-tank.   The tiles need not be more than three feet apart, though twice this distance is not unusual.   In fact, the system is in this respect a very flexible one, and can be adapted to land of any shape or inclination.   The fall of the main line from the tank to the absorption-drains should not, especially after coming within 20 feet of the first line of tiles, have a fall of more than 4 inches to 100 feet.   Its joints should be cemented, and the branches for connection with the tile lines should come out from the bottom of the tile, not from the middle, as is usual with branches of vitrified pipe.   Special pieces for this purpose are to be obtained from the dealers.   The absorption-lines themselves should have a fall of not more than 2 inches per 100 feet: more than this gives a tendency to an accumulation of sewage at the far end of the line, and if the line is long, to a breaking out of the sewage at the surface.

Figures 7, 8, and 9 show three different methods of applying this system according to variations of the ground.   In each case the dotted lines are contour-lines showing differences of elevation of one foot.

In Figure 7, the flush-tank *A* receives its sewage
from the sink-drain leading from a corner of the house.
Its discharge is through a direct line to the point *c*,
starting at the tank at a depth of three or four feet
below the surface and coming to within a foot of the
surface at the point *c*. At the point *c* the main drain
is turned at an angle and has three different outlets to
be used, one at a time, in alternation. The first one
communicates with two parallel drains at the bottom of

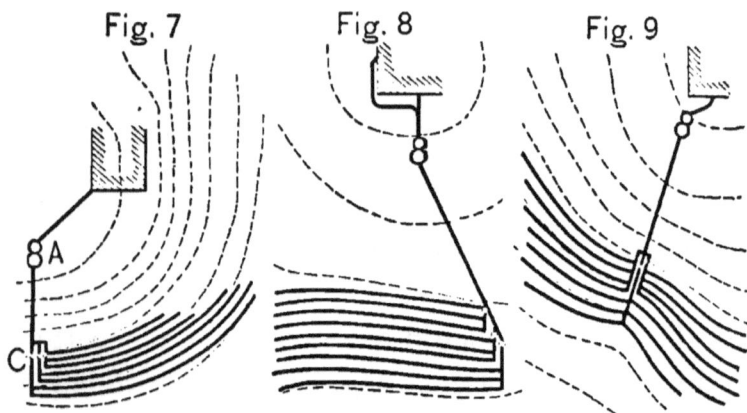

Fig. 7   Fig. 8   Fig. 9

the field, these having a sufficient combined length to
receive the whole discharge of the tank. The second
one runs parallel to the first, to two absorption-drains
corresponding with the first two. The third communicates with the other system of three parallel lines, which
are shorter, having about the same aggregate length as
the two of the other systems. They are carried around
nearly parallel to the contours to secure the requisite
slight fall.

In Figure 8 the flush-tank is fed by two drains from

the house and connects, as shown, with three series of three drains each. The land is much more nearly level, and the nine absorption-drains are in ground having a total fall of only one foot.

In Figure 9 the flush-tank is fed by a single drain, not straight, and connects with its alternating gate. There are two series of drains, four on one side of the field and four on another, while a third line connects with the series of three shorter drains on each side of the medial line.

The flush-tank may be placed in any position and at any distance from the house, and the field may be at any distance from the flush-tank to which a proper fall can be obtained.

One modification of this system which will obviate much of the difficulty by requiring a greatly reduced retention of solid matters, where indeed a coarse screen may, in some cases, be made to hold back all that is necessary to retain, consists in the use of larger absorption-tiles, say 4-inch or 6-inch, with open joints, fully one-half inch wide and laid in coarse gravel or other very open material just under the surface of the ground in two, or better, three series, each of which has sufficient capacity in its pipes to receive the entire contents of the tank at each discharge. The discharge being at intervals of from twelve to twenty-four hours, the liquid sewage with its soluble and its finer suspended impurities will have ample time to leak out into the soil; and during the period of intermission, while the other two series of drains are being used, worms, beetles, and other insects will consume, or decomposition will destroy, the relatively small amount of deposited matter, which, however, might be sufficient to obstruct 2-inch tiles.

A still further modification consists in making the drains of similar large tiles of "horseshoe" shape, laid in a trench filled with coarse gravel or broken stones. The capacity of the tiles and of the voids among the stones in each series should be sufficient to receive more than the full contents of the tank.

Cross-sections of 4-inch horseshoe tiles laid in trenches filled with stone or gravel are shown in Figures 10, 11, and 12.

In Figure 10, the ground is supposed to be reasonably absorptive, like garden mould. In Figure 11, the natural soil is very heavy and non-absorptive, and is level or nearly so. It is thoroughly underdrained and is covered with sand or gravel in low ridges, being deeper at the absorption-drains than midway between them. The purification takes place entirely in this porous and well-aerated surface, the clarified water sinking into the drained ground below. In Figure 12, the land is non-absorptive and has a decided slope. It should be well underdrained with tiles running up and down the slope. The surface is covered with sand or gravel, and is divided into sections by banks of clay (under the sand) at the foot of each section. The sewage is delivered into a horseshoe tile, with broken stone, at the upper side of each section, and is purified (and its water is absorbed) before it reaches the clay bank below, or is held by this bank until it is disposed of.

This question is often raised with reference to the sub-surface system: "If the dire results of the use of

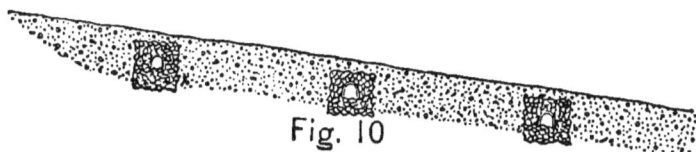

Fig. 10

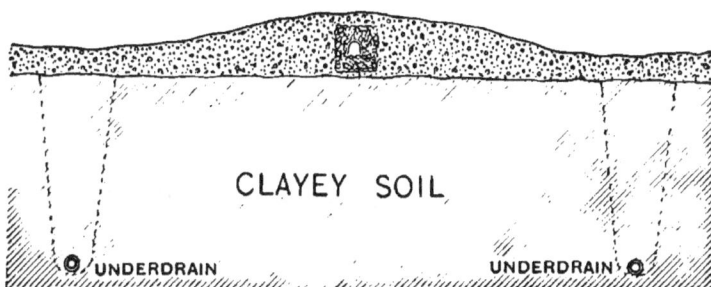

CLAYEY SOIL

UNDERDRAIN   UNDERDRAIN

Fig. 11

Fig. 12

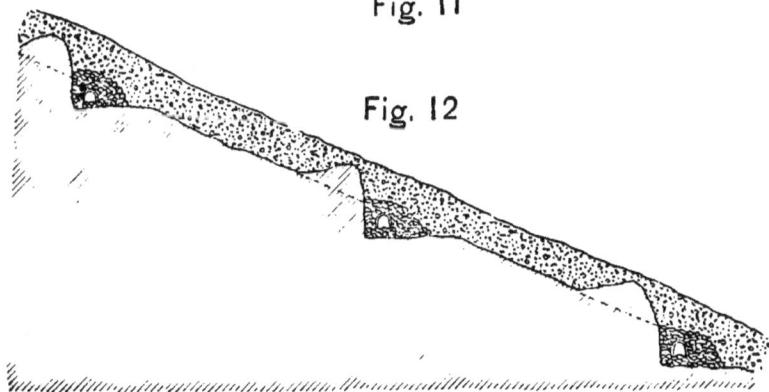

the ordinary leaching cesspool are so serious, why is it not just as bad to discharge the same materials into the ground by leaching drains?" The difference is radical. In the case of the cesspool, the leaching takes place almost entirely at such a considerable distance from the surface that the exclusion of air makes suitable bacterial action impossible, while the delivery through drains laid immediately under the sod is into aerated ground which is teeming with bacteria.

In the system described above, a certain amount of putrefaction is inevitable, and putrefaction is always objectionable. When once its fæcal matter is submerged, so as to prevent the exhalation of its odors into the atmosphere, *fresh* sewage is entirely inoffensive to the smell. It is simply so much dirty water. It might be thrown in considerable quantities on to a grass plat without other objection than that which would attach to the sight of its solid portions. Unless thrown so frequently and in such quantities on the same spot as to saturate the ground, exclude the air, and so prevent proper bacterial action, it would produce no odor whatever, barring the trifling odor of exposed fæces. The same sewage retained for two or three days in a barrel, vault, or cesspool would enter into a state of offensive decomposition, and would constitute a nuisance of a serious character.

The real problem in the disposal of household sewage, and indeed of the sewage of large institutions, and even of towns, is to bring it into contact with a suitable oxidizing medium, while in so fresh a condition as not to be offensive, and to make sure that the drains and other appliances through which it passes shall at no time become offensive. Offensive odor is, of course, chiefly to be guarded against; but offence to the sight

would also constitute an important objection. No
method has as yet been devised by which the whole
process could be appropriately carried on on the front
lawn of a dwelling-house. We have, however, arrived
at a point where, even in a reasonably secluded back-
yard, all of the conditions may be satisfied. Setting
aside exceptional cases, where only the sub-surface
system with a double-chambered flush-tank would be
acceptable, the work may be done in a great majority of
instances by a system of surface disposal or of sub-surface
disposal with large pipes, with an automatic flush-tank
which may at all times be thoroughly aerated and which
may be cleansed with sufficient frequency to prevent
odor from its sliming. Especially in connection with
such a flush-tank can the ventilation of the main drain
and soil pipes be maintained with such completeness
as to prevent the accumulation of odors in them. The
objection to placing a flush-tank of this character quite
near to a house is rather fancied than real.

The nearer it is and the more constantly subjected to
inspection, the greater will be the certainty of keeping
it always in good condition. On the other hand, the
farther from the house, the more will the solid parts of
the sewage be broken up in transit, the less imperative
will be the need for scrupulous cleanliness, and the less
the care required; civilized living, however, requires
great care in all such matters. Civilization is only
lately beginning to take cognizance of this requirement,
and the public at large is still disposed to like best a
system which calls no attention to itself, and of which

the objectionable features are so hidden as to be easily forgotten. Even under their best development, such systems belong to the "out-of-sight-out-of-mind" class.

Until 1879, when the writer constructed for the late E. F. Bowditch, of Framingham, Mass., a water-closet consisting of a white earthenware hopper set in a white tile floor surrounded with white tile walls with nothing in front of it, and with only a seat without a cover over it, this seat being hinged to turn back and expose the hopper, until then it was the universal rule, in housework, to convert the water-closet into an unobtrusive piece of furniture by enclosing it tightly in wood-work — often obtrusive cabinet-work — with a lid attempting to conceal its character. Now, a dozen years later, all "first-class" plumbing supplies only such fully exposed closets, frankly showing what they are and demanding, and receiving, the constant care which removes the least suggestion of offensiveness or indecency.[1] The modern closet so constructed is in every respect unobjectionable; the old closet, depending on concealment for its decency, is almost universally indecent. It manifests itself, not to the sight, which soon becomes accustomed to whatever is needful and appropriate, but to the sense of smell, which never becomes accustomed to the fetid odor of a highly

[1] Plumbers are, recently, making a concession to self-conscious immodesty by adding a hinged cover over the seat. Like hiding the face behind a fan, this invites attention to what is concealed. The only really modest treatment is the perfectly frank treatment; — show the whole apparatus, and show that it is absolutely clean.

ornamental pan-closet, even though built around with polished rosewood.

Unquestionably, the same principle will apply in the case of apparatus for the disposal of sewage outside of the house. The cesspool, the vault, and the double-chambered flush-tank will soon have become things of the past, among those who care for good sanitary conditions. Their places will be taken by some device having the essential features of the system described below, combining efficiency with the possibility of, and the demand for, perfect cleanliness. This being accomplished, the "horrid drains" will soon cease to exist, or to be thought about.

It is not to be supposed that this precise manner of applying the system will long remain the accepted one. But the principle of the arrangement seems to embody all the elements of permanence. This principle may be thus stated: —

Deliver the sewage as soon as produced, through thoroughly flushed and ventilated pipes and drains, to a point outside of the house; hold back its coarser substances by some form of screen which will allow everything to pass that can, in an unobjectionable way, be disposed of with the sewage; accumulate the regular flow through the drain in a receptacle large enough to hold the supply of a few hours, or of a day, as the case may be; the receptacle becoming full, discharge its contents automatically, rapidly, and completely on to the surface of the ground, or into drains immediately below the surface, for its final, complete, and inoffensive disposal;

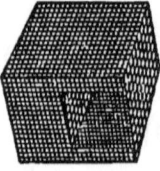

Fig. 16

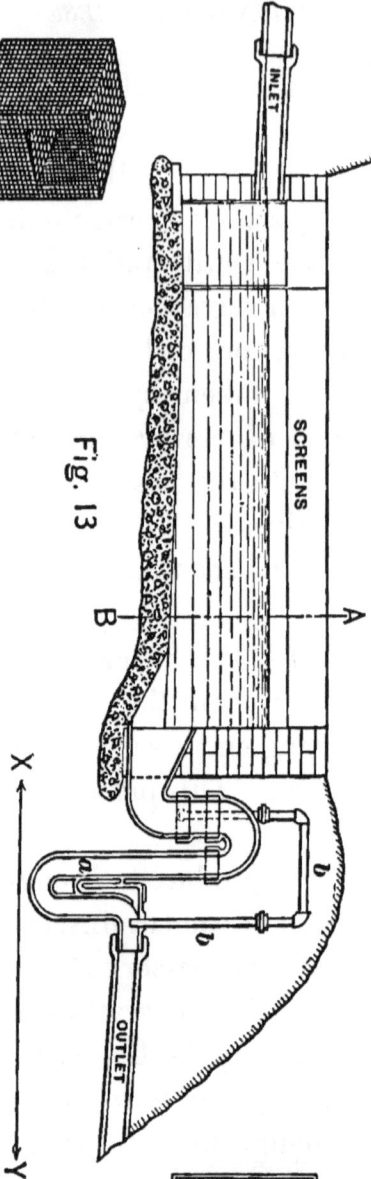

INLET

SCREENS

Fig. 13

B — — — A

X

a

b

b

OUTLET

Y

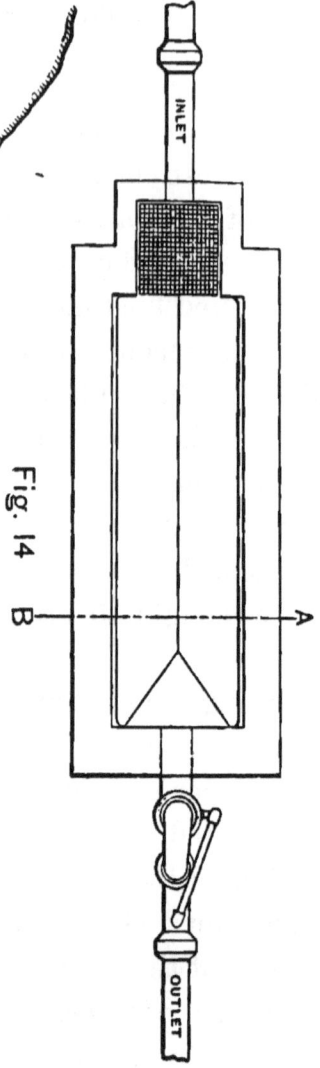

INLET

Fig. 14

B — — — A

OUTLET

Fig. 17

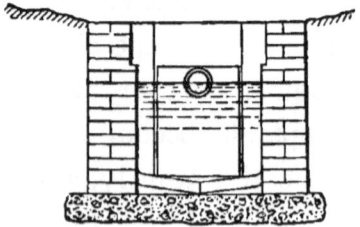

Section A B
Fig. 15

230

arrange the screen and tank in such a way that they may be kept in a cleanly condition with little labor and without requiring the constant supervision and nagging of the master of the house. The appliances shown in the Figures 13 to 20, and the method of working described in connection therewith, seem to constitute a satisfactory application of this principle.

Figure 13 shows a vertical, longitudinal section, Figure 14 a plan, and Figure 15 a vertical cross-section of a new form of flush-tank of which the inside measurements are, a width of one foot eight inches, a length of six feet, and a height of two feet. At a distance of four inches below the top on the long sides, a ledge two inches wide is formed by setting the brickwork of the walls that far back.

This ledge is intended to receive wire-cloth screens twenty-four inches square, shown in Figure 17. The length of the tank may be increased to any number of multiples of two feet in order to obtain the desired capacity. A tank eight feet long would take four screens, a tank ten feet long, five screens, etc. The uniform width is maintained, as the screens are made only in the one size, and as they would be liable to sag if much wider between their supports. The bottom of the tank is so graded as to deliver its contents entirely at the centre of the discharging-end, where there is a depression equal to the receiving-height of the throat of the siphon at its narrowest part. At the end of the tank there is built a recess fifteen inches square to receive the screening-cage shown in Figure 16.[1] This cage is made of galvanized-iron-wire cloth with 1-inch meshes. It is entirely closed at the top and bottom and on three of its sides. One of its sides, that which is to be placed next to the inflowing drain, has an opening at its top ten inches square. This cage constitutes a complete screen to withhold whatever will not pass a 1-inch mesh, — paper and all solids of considerable size. The agitation of its contents by the inflow will break up much of the softer solid parts of the sewage and carry them through the meshes; what will not so pass must be retained, because it would tend to obstruct pipes in the case of sub-surface delivery, and might make objectionable deposits on the ground in the case of surface delivery. These cages are furnished in duplicate, so that whenever one is removed for cleaning, another can be substituted for it immediately. The one removed, after standing a few minutes, will have parted with all of its liquids, and its solid contents can be shaken out through the 10-inch opening and removed, or dug into the ground. When the cage and the covers, Figures 16 and 17, are all in place, the whole tank is sufficiently screened from observation and is protected against leaves and rubbish which might otherwise get access to its contents. As often as experience shows it to be necessary, perhaps daily, the

---

[1] As shown in the illustrations, the sewage enters at one end of the tank and flows out at the other. It will be better to make the inlet and the outlet at the same end, so that deposits forming near the inlet will have the full flow to remove them.

covering-screens, Figure **17**, should be removed, after discharging the tank, and its walls and bottom should be thoroughly swept down, the sewage accumulated in its outlet being sufficient for such washing. As above indicated, the frequency with which this cleansing should be performed may vary according to nearness to, or remoteness from, the house, walks, etc.

Figure **22** shows the masonry construction of this tank, the material being brick glazed on the inner face, and marble or other slabs at the top.

The tank is discharged after its contents reach a certain height by the action of a Rhoads-Williams automatic siphon placed entirely outside of the tank, having a funnel-shaped inlet for the entrance of the sewage. This siphon will require no attention. Whenever the tank fills to the discharging-line, the whole accumulation will flow out rapidly, and when the flow ceases the siphon will "break," allowing no further discharge until the tank has filled again.

The tanks may be built of ordinary brickwork laid and coated inside and out with Portland cement, or with stone or concrete similarly coated. They may be cheaply but simply made, or they may even be lined and capped with white marble. Another excellent material would be white or straw-colored glazed bricks laid with close joints. Elegance of finish will not be altogether useless; for the finer they are the more easily and the more certainly will these tanks be kept in good condition.

In a single case, all others so far heard from working satisfactorily, it was found that the flocculent matter

passing the screen clogged the 4-inch absorption-tiles
after a time. This was obviated by constructing a
settling-chamber, such as is shown at the left end of
Figure 5, in the course of the drain from the house to
the flush-tank. This may be necessary in other cases.
For surface disposal, it has been clearly shown not to
be required.

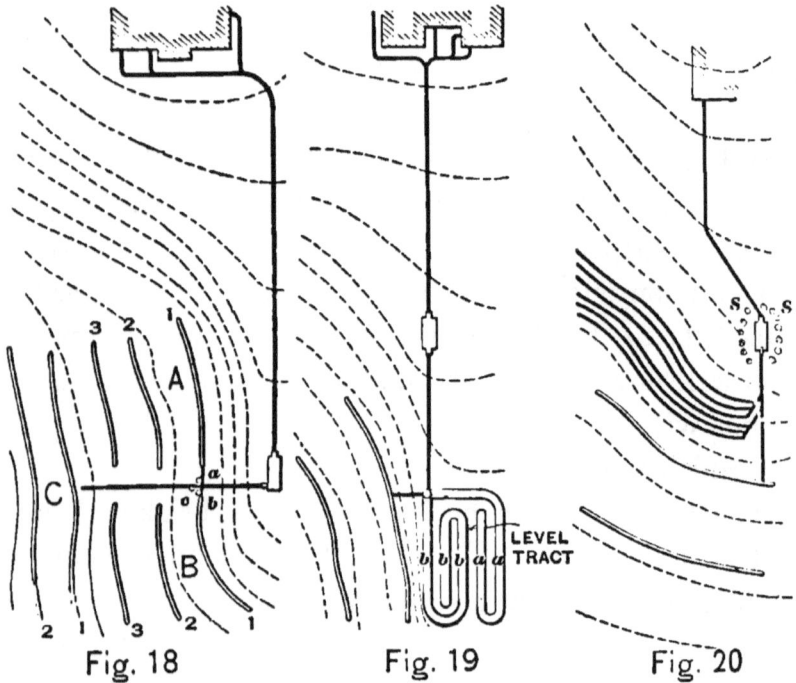

Fig. 18   Fig. 19   Fig. 20

Figure 18 shows the flush-tank illustrated in Figures
13 to 17, placed at some distance from the house,
receiving sewage from three house-drains, and delivering its contents for surface disposal by the use of three
alternating systems of surface gutters or barriers to
equalize the flow. These sections are marked *A*, *B*,

and C, the gates a, b, and c regulating the distribution. In A and B the sewage is delivered along the upper edge of gently sloping land. If the land is steeper, the gutters or barriers must be nearer together to equalize the flow. Water escaping from the upper gutter, or barrier 1, is collected again for a uniform flow at barrier 2, and again at barrier 3. Disposal for section B operates in the same manner. On section C, the gutters or barriers being much longer, only two are needed. These illustrations are not drawn to scale, and are only intended to illustrate the general features of the process. The gutters or barriers must be absolutely horizontal, and so arranged that the sewage escaping from them will flow evenly over the land below. The distribution may require the cutting of a leader-furrow here and there in the grass with a spade.

This method of surface irrigation removes absolutely all impurity from the sewage; what becomes of it after it has passed over a sufficient area of ground is immaterial. If it escapes into the brook or other watercourse, it will by that time have become purer than the water of the brook itself.

Figure 19 shows a system in which the same tank is used, receiving the flow from four house-drains, and delivering its sewage into absolutely level, wide trenches, of sufficient length. In the case shown, there are two of these trenches returned on themselves to give sufficient length. They are marked a, a, a, and b, b, b. In connection with the same system, there is shown a system of surface irrigation on sloping land. The satisfactory use of the trenches *a, a, a,* and *b, b, b,* requires land of very absorptive character, the more porous the better. The best of all is a very fine gravel. As the trenches become filled on the discharge of the flush-tank, the liquid soaking away into the ground, there is left a felt-like coating on the surface, which requires either a sufficient intermission of use to be destroyed by exposure to the air, or what accomplishes the same purpose, a

Fig. 21

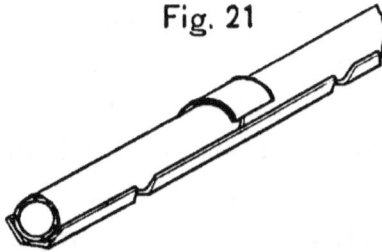

thorough raking of the surface from time to time into the material in which the trenches are cut.

Figure 20 shows the same flush-tank surrounded by a screen of evergreens *s, s;* two systems of sub-surface absorption-drains similar to those shown in Figures 7, 8, and 9; and one system of surface disposal similar to those shown in Figure 18.

The variation of details, such as the size and location of the flush-tank, the arrangement, location, and extent of the surface gutters, or barriers, the horizontal trenches, the sub-surface absorption-drains, etc., may be almost infinite, so that the character of the soil, the formation of the surface, the use to which the land is to be put, the necessity for concealment, etc., may be

accommodated in all cases. The flow from the flush-
tank to the absorption-field is conveniently directed to
the different sections by a simple gate-chamber made
for the purpose.

While gutters cut into the ground will be effective
in collecting and equalizing the flow, they have the
drawback that they retain sewage after the flow ceases

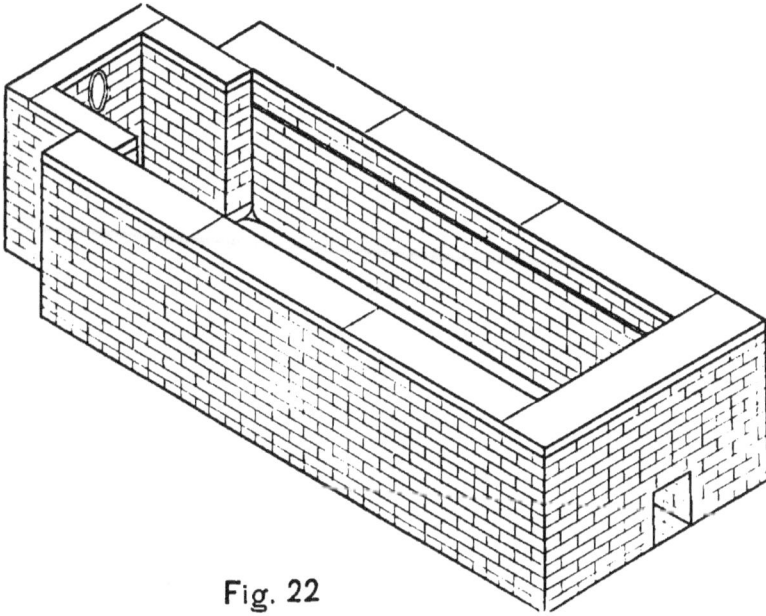

Fig. 22

and become odorous. The porous barriers (of broken
stone, gravel, etc.) allow the whole flow to pass with
only such delay as is needed to equalize the flow. It
is best to lay these barriers on a narrow strip of brick
paving.

It may with advantage be repeated here that while
this system of disposal seems to be as nearly perfect as
is possible in the present state of the art, no such system

will withstand neglect. It affords a perfect solution of one of the most difficult and dangerous problems connected with life in districts where sewers are not available, and the completeness of the result to be secured amply compensates for the slight amount of regular attention required.

CHAPTER XVIII.

CONCLUSION.

It has been the purpose of this work to set forth, in a simple way and in terms as free as possible from technical nomenclature, the principles and the practice of sewage disposal, which have been slowly developed since the time of the publication of the first report of the Health of Towns Commission, in England, fifty years ago (1844).

The purpose has been, not so much to give specific information to engineers, as to inform the public at large, and especially those whose official positions, as members of Boards of Health, and of committees of councils, etc., bring them face to face with the practical sanitary problems which present themselves in all communities, and which become more and more serious as the work of removing liquid wastes by sewers progresses.

There has been, thus far, little popular knowledge of the subject, and consequently little demand for improved methods of disposal. Practical information concerning these has been mainly confined to a few specialists who have given it particular attention, and the most active influence is often exerted by those whose business it is to construct works of a special character, and whose knowledge does not cover the whole range of the subject. These advocates, usually interested in some form of mechanical or chemical treatment, are not the safest guides in the selection of the methods best applicable to varying conditions of particular cases.

What has been chiefly attempted has been to set forth a general review of the whole subject, in terms which will be understood by all intelligent persons who may be called on to investigate the matter, with a view to qualifying themselves for action in regard to it.

The Sewage of Towns Commission (1858) reported a series of conclusions, of which the following are the most important.

"1. That the increasing pollution of the rivers and streams of the country is an evil of national importance, which urgently demands the application of remedial measures; that the discharge of sewage and of the noxious refuse of factories into them is a source of nuisance and danger to health; that it acts injuriously not only on the locality where it occurs, but also on the population of the districts through which the polluted rivers flow; that it poisons the water, which in many cases forms the sole supply of the population for all purposes, including drinking; that it destroys the fish; and generally, that it impairs the value and the natural advantages derived from rivers and streams of water.

"2. That this evil has largely increased with the growing cleanliness and internal improvements of towns as regards water-supply and drainage; that its increase will continue to be in direct proportion to such improvements; and that, as these improvements are as yet very partial, the nuisance of sewage, already very sensibly felt, is extremely slight as compared with what it will become when sewage and drainage works have been carried into full effect.

"3. That in many towns measures for improved water-supply and drainage are retarded, from the difficulties of disposing of the increased sewage which results from them; that the law which regulates the rights of outfall is in an anomalous and undefined condition; that judicial decisions of a conflicting character have been arrived at in different instances, and that consequently the authorities of towns have constantly before them the fear of harassing litigation.

"4. That the methods which have been adopted with the view of dealing with sewage are of two kinds: the one being the application of the whole sewage to land, and the other that of treating it by chemical processes, to separate its most offensive portions; that the direct application of sewage to land favorably situated, if judiciously carried out and confined to a suitable area exclusively grass, is profitable to persons so employing it; that where the conditions are unfavorable, a small payment on the part of the local authorities will restore the balance.

"5. That this method of sewage application, conducted with moderate care, is not productive of nuisance or injury to health.

R

"6. That when circumstances prevent the disposal of sewage by direct application to land, the processes of precipitation will greatly ameliorate, and practically obviate, the evils of sewage outfalls, especially where there are large rivers for the discharge of the liquid; that such methods of treating sewage do not retain more than a small portion of the fertilizing matter, and that although in some cases the sale of the manure may repay the cost of production, they are not likely to be successful as private speculations."

These conclusions are as valuable now as they were when they were written.

The work of arranging for the disposal of the sewage of a town is a most serious one from every point of view, and it involves special knowledge, which cannot be communicated in books, as well as special capacity re-enforced by experience. At the same time, the necessity for such work, and the ability to exercise a proper controlling judgment concerning it, are well within the range of the influence of popular discussion. The field is a wide one. Its development in this country has hardly been begun. We have thousands of communities whose sanitary condition will not be satisfactory until the disposal of their wastes is cared for in a practical and scientific manner; yet Mr. Baker's list, which is as complete as he could make it, includes only about thirty towns, large and small, in the whole United States where disposal works, more or less good, have been established.

The writer ventures to hope that this book may aid

in the awakening of popular interest in the subject, and in the dissemination of popular knowledge concerning the principles by which works of disposal should be controlled. It is by the awakening of such interest and by the spread of such knowledge that the public health is to receive the greatest benefit that it remains to sanitary engineering yet to give it.

INDEX.

(

SEWERAGE AND LAND DRAINAGE.

BY

GEORGE E. WARING, JR., M.I.C.E.,

Fellow of the Sanitary Institute of Great Britain ; Corresponding Member
of the American Institute of Architects.

Quarto, cloth. **With numerous Illustrations, many colored. $6.00.**

"The whole of the discussion is of great value to experts and to public officers charged with providing sewerage, who under our system cannot possibly be experts, but who may, nevertheless, have enough sense of official responsibility to wish to inform themselves. Such persons will find no difficulty in following the argument here given, for one advantage of the newness of the science is that it has not yet accumulated a technical jargon, and the book is composed throughout in the English language." — *New York Times.*

"This volume, which might have been a dry and technical discussion, fit only for engineers, is a clear and interesting account of its title subject. Colonel Waring's facility of expression has presented a popular view without losing sight of the serious facts. In part it is an appeal to the laity, but not at the sacrifice of honesty. . . . While not endorsing as unimpeachable every proposition advanced, we do frankly advise that this may be taken as general authority over the wide range it covers, and as such be admitted to an honorable place in every public library, and as a reference book for sanitary engineers, for officers of health, and intelligent citizens." — *The Nation.*

"We have before us one of the most elaborate and important additions to the literature of sanitary engineering published. The advancement of sanitary science has come to embrace the labors, in a large part, of the domestic and civil engineer, and as information on this vital subject is widely demanded, this work is timely, and cannot fail to have a large clientage. The reputation of Mr. Waring has far preceded his work, and he has attained eminence in two continents through his efficient labors and former publications. . . . The whole subject of domestic engineering is fully treated, and will prove not only valuable to the profession, but instructive to the general public. It is a text-book on sanitary engineering to be studied and kept convenient for reference." — *The Sanitary News.*

"Colonel Waring's book is in his well-known clear, concise, and forcible style, and the publishers have nothing failed in their art to make it attractive. Physicians will find much in it of importance to them, particularly the chapter on house-drainage. In towns planning sewerage it is well-nigh indispensable." — *Boston Medical and Surgical Journal.*

"None acquainted with the subject need to be reminded that Colonel Waring is not only a pioneer in the development of sanitary science in this country, but has kept himself steadily abreast of the foremost in its practice." — *New York Commercial Advertiser.*

D. VAN NOSTRAND COMPANY,

23 MURRAY AND 27 WARREN STREETS, NEW YORK.

SEWAGE DISPOSAL IN THE UNITED STATES.

BY

GEORGE W. RAFTER, M. Am. Soc. C.E.,

AND

M. N. BAKER, Ph.B.,

Associate Editor, "Engineering News."

8vo, cloth. With numerous Illustrations. $6.00.

"This is the most elaborate and, for engineers, the most useful book that has appeared on the important subject of which it treats. In fact, so far as practice in this country is concerned, it is the only comprehensive one. It will surprise even those who are familiar with the subject to know that so much material could have been gathered from actual experience with sewage disposal works carried out in the United States, where the demand for any other disposal than discharge into a stream or into the sea first arose less than twenty years ago." — *The Nation.*

"Altogether it is the most comprehensive and the best work on sewage disposal hitherto published in the United States, eminently useful to hydraulic engineers not only, but to all state boards of health." — *The Sanitarian.*

"Perhaps the most valuable portion of the work to the general public is that treating of the danger arising from the polluting of water courses and water supplies by sewage. Every aspect of the sewage question is treated with the utmost care, with references to the best available authorities." — *Rochester Democrat and Chronicle.*

"It is a well-informed, scientific, and valuable publication." — *New York Times.*

"In eight appendices all present laws bearing on the subject are recapitulated, as enacted. When we say that all these themes are treated in an interesting manner, it is not too much to claim that in addition to bestowing a boon on their own guild, Messrs. Rafter and Baker have conferred a benefit on the public in making this book. The reading of it would transform into an intelligent sanitarian any one who really desires to learn." — *The Independent.*

D. VAN NOSTRAND COMPANY,

23 MURRAY AND 27 WARREN STREETS, NEW YORK.

2

LIST OF BOOKS

ON

SANITARY SCIENCE, SEWERAGE, ᴬᴺᴰ DRAINAGE,

FOR SALE BY

D. VAN NOSTRAND COMPANY, ·

23 Murray and 27 Warren Streets, New York.

ADAMS, J. W. Sewers and Drains for Populous Districts. 8vo, cloth, illustrated, 1892 . . . . $2.50

BALCH, L. Manual for Boards of Health and Health Officers. 12mo, cloth, 1893   1.50

BAUMEISTER, R. Cleaning and Sewerage of Cities. 8vo, cloth, 1891 . . 2.50

BEARDMORE, W. L. The Drainage of Habitable Buildings. 8vo, cloth, 1892   1.50

BLYTH, A. W. Lectures on Sanitary Law. 8vo, cloth, 1893 . . . . 2.50

—— Manual of Public Health. 8vo, cloth, illustrated, 1890 . . . . 5.25

—— Dictionary of Hygiene and Public Health, comprising sanitary chemistry, engineering, and legislation, etc. 8vo, cloth, illustrated, 1876 . . . . 10.00

BOULNOIS, H. P. Municipal and Sanitary Engineer's Hand-book. Second edition. 8vo, cloth, illustrated, 1892 . . . . 6.00

BROWN, GLENN. Healthy Foundations for Houses. 18mo, boards, illustrated, 1885 50

BROWN, G. P. Sewer Gas and its Dangers. 16mo, cloth, 1881 . . . . 1.25

BURKE, U. R. Hand-book of Sewerage Utilization. 8vo, cloth, 1873 . . 1.40

COLYER, F. Treatise on Water Supply, Drainage, and Sanitary Appliances of Residences. 12mo, cloth, 1889 . . . . 1.50

—— Public Institutions: Engineering, Sanitary and Other Appliances. 8vo, cloth, 1889 . . . . 4.20

CORFIELD, W. H. The Treatment and Utilization of Sewage. Third edition. 8vo, cloth, 1887 . . . . 4.50

—— Water and Water Supply. 18mo, boards 50

—— Sewerage and Sewage Utilization. 18mo, boards, 1875 50

—— Dwelling Houses; their Sanitary Construction and Arrangements. 18mo, boards, 1880 50

CRIMP, W. S. Sewage Disposal Works. 8vo, cloth, 1890 . . . . 7.50

—— Sewage Treatment and Sludge Disposal. 8vo, sewed, 1893 60

DEMPSEY, G. D., and CLARK, D. K. On the Drainage of Lands, Towns, and Buildings. Second edition, revised. 12mo, cloth, illustrated, 1890 . . 1.80

DENTON, J. B. Sewage Disposal. 8vo, cloth, 1881 . . . . . . 1.40

FANNING, J. T. Practical Treatise on Water-supply Engineering. 8vo, cloth, illustrated 5.00

FOLKHARD, C. W. Potable Water and the Different Methods of Detecting Impurities. 18mo, boards, 1882 50

GERHARD, W. P. Recent Practice in the Sanitary Drainage of Buildings. Second edition. 18mo, boards, 1890. 50

—— Disposal of Household Waste. 18mo, boards, 1890 50

—— House Drainage and Sanitary Plumbing. Sixth edition. 18mo, boards, illustrated, 1894 50

—— Guide to Sanitary House Inspection. Third edition. Square 16mo, cloth, 1890 . . . . 1.00

HELLYER, S. S. The Plumber and Sanitary Houses. Fifth edition. 8vo, cloth, 1893 . . . . 5.00

3

LIST OF BOOKS ON SANITARY SCIENCE, SEWERAGE, AND DRAINAGE.

KENWOOD, H. R. Public Health Laboratory Work. 12mo, cloth, illustrated, 1893 . . . . $3.00

MAGUIRE, W. R. Domestic Sanitary Drainage and Plumbing, Lectures on Practical Sanitation. 8vo, cloth, illustrated, 1890 . . . . . . 4.50

NICHOLS, W. R. Water Supply, Considered mainly from a Chemical and Sanitary Standpoint. Fourth edition. 8vo, cloth, illustrated, 1892 . . . 2.50

PALMBERG, A. Treatise on Public Health and its Applications in Different European Countries. Translated from the French by A. Newsholme. 8vo, cloth, illustrated, 1893 . . . . 5.00

PARKER, L., and WORTHINGTON, R. H. The Law of Public Health and Safety, and the Powers and Duties of Boards of Health. 8vo, sheep, 1892 . 5.25

PARKES, L. C. Hygiene and Public Health. Third edition. 8vo, cloth, illustrated, 1892 . . . . 2.75

POORE, G. V. Essays on Rural Hygiene. 12mo, cloth, 1893 . . . . 2.00

RAFTER, G. W. The Microscopical Examination of Potable Water. With diagrams. 18mo, boards 50

RAWLINSON, R. The Public Health. Suggestions as to the Preparation of District Maps and Plans for Main Sewerage, Drainage, and Water Supply. Folio, paper, illustrated, 1878 . . . . 1.20

REEVES, R. H. Sewer Ventilation and Sewage Treatment. 12mo, cloth, 9 folding plates, 1889 . . . . 1.40

ROBINSON, H. Sewage Disposal ; containing Information for Sanitary Authorities and Sanitary Engineers. Second edition. 12mo, cloth, 1882 . . . 2.00

SEWAGE PURIFICATION IN AMERICA. A Description of the Municipal Sewerage Purification Plants in the United States and Canada. 12mo, paper, illustrated, 1893. . . . . 1.00

SIMON, SIR J. English Sanitary Institutions. 8vo, cloth, 1890 . . . 8.00

SLAGG, C. Sanitary Work in the Smaller Towns and Villages. Revised edition. 12mo, cloth, 1893 . . . . 1.40

SLATER, J. W. Sewage Treatment, Purification, and Utilization. 12mo, cloth, 1887 . . . . 2.25

SMEATON, J. Plumbing, Drainage, Water Supply, etc. 8vo, cloth, illustrated, 1893 . . . . 3.00

STALEY, C., and PIERSON, G. S. The Separate System of Sewerage. Its Theory and Construction. Second edition. 8vo, cloth, 1891 . . . . 3.00

SYKES, J. F. J. Public Health Problems. 12mo, cloth, illustrations and maps, 1892 . . . . 1.25

TAYLOR, A. Sanitary Inspector's Hand-book. 12mo, cloth, illustrated, 1893 . 2.00

TIDY, C. M. The Treatment of Sewage. 18mo, boards, 1887 50

VARONA, A. de. Sewer Gases, their Nature and Origin. 18mo, boards, 1881 . .50

WARING, G. E. Sewerage and Land Drainage. Third edition. 4to, cloth, illustrated, colored plates, 1891 . . . . 6.00

—— Sanitary Condition of City and Country Dwelling Houses. 18mo, boards, 1877 50

—— The Sanitary Drainage of Houses and Towns. 12mo, cloth, 1879 . . 2.00

—— How to Drain a House. Practical Information for Householders. 12mo, cloth, 1895 . . . . 1.25

—— Modern Methods of Sewage Disposal for Towns, Public Institutions, and Isolated Houses 2.00

WILLOUGHBY, E. F. Hand-book of Public Health and Demography. 16mo, cloth, 1893 . . . . 1.50

—— Health Officer's Pocket-book. A Guide to Sanitary Practice and Law for Medical Officers of Health, Sanitary Inspectors, etc. 16mo, cloth, illustrated, 1893 . . . . 3.00